The Way of the Cell

The Way of the Cell

Molecules, Organisms and the Order of Life

Franklin M. Harold

OXFORD
UNIVERSITY PRESS
2001

OXFORD

UNIVERSITY PRESS

Oxford   New York

Athens   Auckland   Bangkok   Bogotá   Buenos Aires   Cape Town
Dar es Salaam   Delhi   Florence   Hong Kong   Istanbul   Karachi
Kolkata   Kuala Lumpur   Madras   Madrid   Melbourne   Mexico City
Nairobi   Paris   Shanghai   Singapore   Taipei   Tokyo   Toronto   Warsaw

and associated companies in
Berlin   Ibadan

Copyright © 2001 by Franklin M. Harold

Published by Oxford University Press, Inc.
198 Madison Avenue, New York, New York 10016

Oxford is a registered trademark of Oxford University Press

Library of Congress Cataloging-in-Publication Data

Harold, Franklin M.
  The way of the cell : molecules, organisms and the order of life / by Franklin M. Harold.
    p. cm.
  Includes bibliographical references and index.
  ISBN 0-19-513512-1
  1. Cytology—Popular works.   2. Life (Biology)—Popular works.   I. Title.

QH582.4 .H37 2001
571.6—dc21   00-056670

1 3 5 7 9 8 6 4 2

Printed in the United States of America
on acid-free paper

ABOUT THE AUTHOR

Frank Harold was born in Germany, grew up in the Middle East and was educated at the City College of New York and the University of California at Berkeley. His professional career spans forty years of research and teaching, mostly in Colorado; he is presently Professor Emeritus of biochemistry at Colorado State University. Dr. Harold's scientific interests center on the physiology, energetics and morphogenesis of microorganisms, and he is a member of the American Academy of Microbiology. He is also a keen traveler and outdoorsman, and a lifelong student of Asian history and civilizations.

"A momentous change had come about when what scientists did came to be taken for granted, even by those who understand little or nothing of it. The crucial change in the making of the modern mind was the widespread acceptance of the idea that the world is essentially rational and explicable, though very wonderful and complicated."

John M. Roberts
The Triumph of the West, 1986

For Ruth: Microbiologist, artist, traveler, hill walker, friend, colleague, wife, mother; and the best thing that ever happened to me.

CONTENTS

PREFACE

This book is not about biology, biochemistry or any other finished and finite discipline, but about life. Life seems to me the supreme marvel of the universe—familiar, thoroughly material, probably ubiquitous yet elusive and ultimately mysterious. My purpose is to assess how far we have come toward a scientific understanding of the phenomenon of life.

With so broad, not to say nebulous a subject, it seems best to spell out the premises on which this inquiry rests. First, I am a scientist by profession, not a philosopher; we shall be concerned here with what natural science has to say about the nature of life, not how it appears to a psychologist, theologian, poet or epistemologist. Second, I take it that the term "life" designates a real phenomenon, recognizable by a set of properties characteristic of some natural objects and lacking in others; one of our goals must be to identify the essential features that distinguish living organisms from other things. Although we have been able to study but one kind of life, the terrestrial variety, it is likely that life exists elsewhere in the universe, and it is arguable that life everywhere will be based on this common set of general principles. Third, during the past century we have come a very long way by scrutinizing the workings, architecture and chemistry of cells and organisms; what we have learned makes a solid foundation for reflection on the nature of life in general. Finally, I hold that the quest for an answer to the riddle, "What is Life?" is one of the grand themes that resonate through the scientific conversation of this century—a period whose science is also its singular glory. That riddle embraces and transcends the subject matter of all the biological sciences, and much of physical science as well. A physics that has no place for life is as impoverished as would be a biology not informed by chemistry. The study of life as a natural phenomenon, a fundamental feature of the universe, must not be allowed to slip into the black hole of departmental tribalism.

Let me enlarge for a moment on the latter point, for herein lies much of the motivation for writing this book. What science knows of the nature of life, it owes to the labors of countless specialists—physicists and chemists, mathematicians and geologists, geneticists and biochemists and physiologists, biologists evolutionary and biologists molecular. The fruits of our labors are first inscribed in shelf upon shelf of professional journals, and subsequently reincarnated in textbooks that have grown too heavy to carry, let alone read. But the nature of life is not a practical topic for research. General insights, if there be any, must be distilled from numberless particular discoveries, and here time may no longer be on our side. The relentless accumulation of information on all subjects, however desirable in itself, frustrates understanding by pressing everyone into ever narrower borders. A second hindrance is the spirit of the times, the clamor that knowledge has value only insofar as it lends itself to practical ends. Scientists themselves increasingly subscribe to the thesis that science must serve the uses of power, not of philosophy: it is, after all, on our usefulness that we base our claim to scarce public resources. The most productive era of fundamental inquiry may thus be approaching an end, and that makes it timely to gather the threads of knowledge spun out by research and see what pattern they make.

That conversational word "understanding" has already cropped up several times, and since it stands for the object of this entire exercise some attempt at definition is in order. Scientists use the word in a somewhat special sense, that was nicely set forth by Mary Midgley in her critical study, *Science as Salvation*. "Understanding anything is finding order in it. . . . It is simply putting [the pattern] into the class of things meaningful—noting how its parts relate to it as a whole, and how it itself relates to the larger scene around it" (1). And "explanation?" To strive for a plausible, self-consistent view of the world, and to communicate it to others, is a less exalted quest. But I cannot agree with those who dismiss it as unworthy, as long as we remember the difference.

One response to the question, What is Life? is simply, Look Around! Note the birds and the butterflies, zebras and ammonites, the intricate web of life present and past, and join the unending struggle to ensure its continuance in the face of human arrogance and mindlessness. This has been eloquently said by others, far better than I could, and it is not what I have in mind here. For the past forty years, I have been immersed in research on the biochemistry and physiology of microorganisms, with emphasis on the fundamental aspects such as bioenergetics and morphogenesis. In consequence, the central problems of life present themselves to me at the interface of chemistry and biology. How do lifeless chemicals come together to produce those exquisitely ordered structures that we call organisms? How can molecular interactions account for their behavior, growth, reproduction? How did organisms and their constituents arise on an earth that had neither, and then diversify into the cornucopia of creatures that enliven each drop of pond water? My purpose is not to "reduce" biology to chemistry and physics, but to gain some insight into the nature of biological order. In an earlier book, I wrote that "Living things differ from non-living ones most pointedly in their capacity to maintain, reproduce and multiply states of matter characterized by an extreme degree of organization" (2). This still rings true; biological organization is the key to the nature of life, and the central theme of this book.

Most branches of science bear on the problem of life, but some are more pertinent than others. This book celebrates microorganisms, and that requires explanation because with most folks the word "life" does not conjure up the image of bacteria or protozoa. Microorganisms do not receive much attention in books about biology, and the public at large knows them chiefly as agents of disease. We who love microbes are apt to justify our peculiar passion by extolling their diversity, far greater than that of all higher forms of life. We point to their manifold interactions with humans, more commonly beneficial than harmful; and we insist that the operation of the biosphere is wholly dependent on microbial ministration. But the reason they star in this book is that reflection on the smallest and simplest forms of life has been singularly fruitful: whether it is molecules we seek to understand or organisms, communities or evolution, the proper study of mankind is often not man but microbes. Geological history reinforces the point. Microorganisms, the bacteria and protists, can make a biosphere all by themselves, and did so for billions of years when the earth was young. Higher organisms hold mysteries that are of special concern to us humans: the genetic basis of disease, the immune response, embryonic development and the nature of mind are now at the forefront of the research effort. But for the purposes of an inquiry into the nature of life, these are peripheral issues. They represent potentialities inherent in living matter, but are not required for its existence.

A company of potential readers looked over my shoulder throughout this writing. First my colleagues from academe, quick to find fault, demanding factual accuracy at all costs and restraining my penchant for generalization; I have heeded their admonitions as best I could. But there was also a party of students, surfeited with facts but curious still, wanting to be reminded of what it was that drew them into science in the first place. A couple of teachers came from a local college or high school, charged with making science intelligible to students whose interests all lie elsewhere; at the end of the day, such teachers may hold in their hands the future of science. I spotted a few members of that endangered species, the educated laity, pleading for simplicity and clarity in the face of unavoidable complexity. And every so often Lyndon Johnson's ghost would whisper in my ear the late president's favorite question: And therefore, what? The latter all persuaded me to step back from technicalities and detail, and to paint with a broad brush. In the end I tried to write the kind of synoptic and non-technical volume that I myself seek out when my reading draws me into strange waters.

Inevitably, then, this is a personal book—one scientist's attempt to wring understanding from the tide of knowledge. It grew out of the experience of a lifetime devoted to research, scholarship and instruction; but since my purpose is to make sense of the facts of life rather than to expound the facts themselves, this inquiry walks the edge of science proper. The arguments and conclusions presented here seem to me sound, but they are certainly not the last word on the subject. The most valuable lessons that the discipline of science teaches are to play the game of conjecture and refutation, to appreciate the provisional nature of our knowledge, and to prize the doubt! If what I have written here encourages a few readers to look up from their gels and genes to peer at the far horizon, I shall be well content. Of my shortcomings as an investigator, scholar, philosopher, and expositor I am keenly aware. But I can claim to share one merit with Erwin Schrödinger, who gave us our marching orders fifty years ago: I, also, am willing to make a fool of myself in a good cause.

ACKNOWLEDGMENTS

No man is an island, a scholar least of all. This book owes something to most everyone that I have ever engaged in serious conversation about life, the universe and the nature of science. They are too numerous to mention by name, even if I could call them to mind, but I must not fail to acknowledge those among my friends and critics that made me think again. Salomon Bartnicki-Garcia, Dennis Bray, David Deamer, Ford Doolittle, Marty Dworkin, Joseph Frankel, Brian Goodwin, Ruth Harold, Lionel Jaffe, Arthur Koch, Darryl Kropf, Nick Money, Harold Morowitz, Gary Olsen, Norman Pace, Martin Pato, Howard Rickenberg, Willie Schreurs, Mitchell Sogin, Bruce Weber and Carl Woese, thank you all for making the labor of science so great a pleasure. Colleagues in my present department tolerated my eccentricities and created an environment in which I could remain productive past formal retirement. James Bamburg, Norman Curthoys, Jenny Nyborg, Marvin Paule, Craig Schenk, Robert and A-Young Woody, thank you. Much of what I have written here was inspired by the teachings of the late Peter Mitchell and Roger Stanier; I believe that they would have approved of this final jog in my professional journey.

My debt extends to many scientists, writers and scholars whom I have never met, but whose books and articles shaped my thinking. I shall cite the most pertinent ones below, but such casual courtesy does not do justice to the influence exerted by John Tyler Bonner, Jacob Bronowski, Paul Davies, Stephen Jay Gould, François Jacob, Lynn Margulis, Ernst Mayr and Rupert Riedl. Special thanks are due to the late Loren Eiseley, from whom I learned during a crisis of doubt that the true use of science is to make the world intelligible. There are also institutions to thank, notably the National Science Foundation, which supported the work of my laboratory as well as the early stages of this writing. Colorado State University did its part, not only by providing support and a library, but also by requiring me to face a class of non-science majors enrolled in Biochem 103 (Cells, Genes and Molecules).

Transformation of a manuscript into a book is again the work of many hands. Nancy Graham turned yellow pages thickly scribbled in longhand into neat, legible typescript, not once but time and again. Individual chapters were scrutinized by D. Bray, D. Deamer, M. Dworkin, R. Harold, D. Kropf, N. Money, G. Olsen, M. Pato, R. Woody and several anonymous reviewers, who spotted more errors and ambiguities than I care to admit. Kirk Jensen proved himself a shrewd and patient editor, whose sage advice let me avoid more than one pratfall. The illustrations were drawn by Gary Raham and Steve McMath at Visible Productions, Inc., Fort Collins, Colorado.

The ever-swelling technical literature presents special problems for one who writes for a general audience. Scholars are expected to acknowledge their sources, but readers find a drizzle of references distracting. By way of compromise, I have tried once more to steer between glibness and pedantry. Most of the citations identify review articles and books, supplemented where necessary by reference to the most recent experimental reports. They are intended to offer a foothold to readers who wish to pursue some aspect in greater depth, but they are far from exhaustive and do not necessarily indicate priority of discovery. Responses to the literature extend through 1999.

And so, having come to the end of a road that began nearly a decade ago, I wonder still what makes one persevere in such a laborious and painstaking effort, whose chief reward is likely to be the satisfaction of bringing it to completion. Is it merely the urge to relieve what the Romans called the writer's itch? The best explanation I have ever found comes from the preface to *Practicing History*, a collection of essays by the eminent historian Barbara Tuchman (1). Let me quote her here for the benefit of anyone who may be contemplating a major writing commitment with mingled dread and desire. "Research is endlessly seductive, writing is hard work. One has to sit down in that chair and think and transform thought into readable, conservative, interesting sentences that both make sense and make the reader turn the page. It is laborious, slow, often painful, sometimes agony. It means rearrangement, revision, adding, cutting, rewriting. But it brings a sense of excitement, almost of rapture, a moment on Olympus. In short, it is an act of creation."

1

Schrödinger's Riddle

"It is better to know some of the questions than all of the answers."

James Thurber

In the spring of 1938, Adolf Hitler launched his conquest of Europe by annexing neighboring Austria. Nazism had struck deep roots in Austria and the populace cheered as the German troops marched into Vienna, but what was left of the country's cultural and intellectual elite scattered into flight. Among those who had left their escape to the last was Erwin Schrödinger, one of the pioneers of quantum mechanics and Austria's premier physicist. Schrödinger found a haven at the Institute for Advanced Studies of Trinity College, Dublin, where his contract required him to deliver a series of public lectures. He elected to discourse on a physicist's view of life and published the lectures in 1944 in the form of a small sprightly volume boldly entitled *What is Life?* (1). It proved to be an enormously influential work that drew students and young scientists into a new biology. I read it as an undergraduate, understood the easier parts and remember that first encounter nearly half a century later.

Rereading Schrödinger today, with the benefit of knowing what came after, leaves one wondering just where the book's appeal lay. Max Perutz, disappointed by the deficiencies of its scientific content, commented acidly that "what was true in his book was not original, and most of what was original was known not to be true even when it was written" (1). Quite so, but the book was directed to a general audience, and it succeeded admirably as a manifesto for a new era. It centered on two topics that were to dominate research for the next thirty years, the nature of the gene and the energetics of biological order; Schrödinger had drawn up an agenda for the new biology. Besides, the book's title posed the crucial question. To be sure, the nature of life is not a subject that experimental science can tackle head on, but it is one that has engaged and eluded natural philosophers for millennia. Schrödinger placed the nature of the gene at the very heart of the mystery and argued that heredity and biological reproduction, which seemed to defy the known laws of physics, could be accommodated within a broader framework. Unlike most physical principles, which were derived by averaging the behavior of large numbers of particles, heredity must reflect the unique properties of one or two individual large molecules. Schrödinger likened these to "aperiodic crystals," which can contain a "codescript" because each group of atoms plays an individual role that is not exactly equivalent to that of any other group. The idea, though not the language, had originally been put forward by a young physicist named Max Delbrück, and one of Schrödinger's objectives was to give Delbrück's insight a wider hearing. "Aperiodic crystal" and "codescript" foreshadowed the structure and function of DNA, the molecule that encodes hereditary information, whose central role had not yet been recognized. But the time was ripe. With the war over, numbers of young scientists (by no means all physicists) were eager to turn their talents to nobler uses, and they responded joyfully to Schrödinger's challenge to bring the study of living organisms fully within the compass of physics and chemistry. The premise that life, though complicated, is rational and explicable has taken root, justified by the tremendous success of Schrödinger's program.

Schrödinger wrote at the beginning of an extraordinary era in biological science, a great eruption of knowledge that cast a brilliant light into the chemical and physical foundations of life. Perhaps one must have lived through this revolution to appreciate how radically it transformed our perception of what biology is about. In the forties, biology was still primarily centered on living creatures, and quite separate from the physical sciences. Research on the molecular constituents was just gathering steam, and the nature of macromolecules in particular was a matter for doubt and debate. Thirty years later, biology seemed well on the way to becoming a province of chemistry. By 1975, it was becoming a tedious but routine task to determine the primary structure of macromolecules, and three-dimensional images were appearing regularly. The pathways by which the major biological molecules are produced and broken down had been worked out. Enzymes continued to challenge the chemical imagination, but how they ensure the high rate and precision of biochemical processes had in principle been clarified. In principle, though not yet in detail, biochemists had discovered how living organisms capture energy and harness it to the performance of work. But the single most spectacular accomplishment was the solution to the problem that Schrödinger had held up as central, the nature of the gene. It led quickly to the discovery of the principles that govern the replication, transmission and expression of genetic information. Those were heady days, splendidly recreated by H. F. Judson in *The Eighth Day of Creation* (2). Unsolved problems remain in all these areas, even today, but we see them as puzzles, not as mysteries. How matter, energy, and information flow through living organisms is nowadays quite thoroughly understood; the ponderous textbooks in which this knowledge is recorded stand as monuments to an achievement that has few parallels in the annals of science.

Can we say then, that the riddle of life has been read—or soon will be, pending only the clarification of a few outstanding details? Those who believe that the object of the quest is to discover the physical and chemical mechanisms that underlie universal biological processes may be inclined to nod assent. But anyone familiar with living creatures will protest that the compendium of molecules and mechanisms omits the very singularities that answer to one's intuitive sense of what "life" means. Surely, a satisfying reading of Schrödinger's riddle should have something to say about cells, those universal units in which the phenomenon of life is dispensed. And it should bear on the kind of observations with which biology has traditionally been concerned: morphology, structure serving function, goal-seeking behavior, reproduction, adaptation. It must, in short, take cognizance of *organisms* in all their complexity, uniqueness and diversity. Physicists and chemists have every reason to take pride in their achievement. But traditional biologists are equally justified in pointing out how much remains to be accounted for, and to wonder whether, in abandoning the organism for its molecules, we have forgotten what the question was.

The open questions about the workings, behavior and functions of organisms differ in degree, and possibly in kind, from the problems that were so satisfyingly solved in the salad days of molecular biology. Biochemical mechanisms, the structure of DNA, even the replication and translation of genetic information are simple—not in the sense of being easy to discover, but in that they involve a limited number of interactions. Moreover, the structures and interactions are literally linear: DNA makes RNA makes protein describes the transformation of one linear set of symbols into another. But when we inquire how an amoeba crawls, or how a yeast cell grows and buds off a daughter, the phenomena are inherently very much more complex. The functions of the living organism typically depend upon the coherent operations of molecules by the million, belonging to hundreds or even thousands of different kinds, and marshalled into order by a hierarchy of controls. Few of these molecules are free in solution. On the contrary, many are first assembled into elaborate constructs whose dimensions are measured in micrometers or even millimeters, orders of magnitude greater than those of individual molecules, and their collective actions characteristically display a direction in space. These features underscore what Warren Weaver, in another seminal essay of the forties (3), called the problems of organized complexity. A satisfying reading of life's riddle demands a rational account of biological organization, and that has yet to be achieved.

During the next fifty years, Weaver thought, science will have to address such questions as "What makes a primrose open when it does?" And we are doing that. We have ample reason to believe that every biological phenomenon, however complex, is ultimately based on chemical and physical interactions among molecules. With this principle as the point of departure, intense efforts are presently underway to understand how and when a flower blooms (and how the amoeba crawls and the yeast buds), by identifying all the relevant molecules and describing how they intermesh. Many of these projects have been successful, some dazzlingly so, and that enables us to supply a mechanistic explanation for a growing number of biological phenomena. Muscle contraction is a case in point: We can explain in a full and satisfying way how the machinery works, given that the molecular elements have been placed in the structural framework revealed by microscopy. But it is not at all clear that we can answer Weaver's question by extrapolating from the molecular parts to the functional whole. If we knew the chemical structure of every muscle molecule, and understood their chemical interactions, would that suffice to specify how these molecules are articulated in time and in space to generate a working muscle?

This is actually a genuine philosophical puzzle, one version of the question whether biology can ultimately be "reduced" to chemistry and physics or is an autonomous science with principles of its own. I shall return to this issue more than once in subsequent chapters, but for present purposes the answer is plainly that there is more to life than just molecular mechanics (4). From the chemistry of macromolecules and the reactions that they catalyze, little can be inferred regarding their articulation into physiological functions at the cellular level, and nothing whatever can be said regarding the form or development of those cells. It therefore seems to me self-evident that the quest for the nature of life cannot be conducted exclusively on the biochemist's horizon. We must also inquire how molecules are organized into larger structures, how direction and function and form arise, and how parts are integrated into wholes. Besides, we must never forget that molecules, cells and organisms are all creatures of history, brought forth by the interplay of chance and necessity. There can be no simple answer to the question, What is Life? It is an invitation to explore the successive levels of biological reality, and a lecture on molecular biology is intrinsically no more (and no less) illuminating than a walk through the woods in the springtime.

Erwin Chargaff made the same point years ago, in one of the most perceptive (and disturbing) autobiographies composed by a scientist (5). "Our understanding of the world is built up of innumerable layers. Each layer is worth exploring, as long as we do not forget that it is one of many. Knowing all there is to know about one layer—a most unlikely event—would not teach us much about the rest." Next time you fly, reflect upon the airplane's wing. It is designed to provide lift, and its component parts serve that function; a skilled mechanic, supplied with aluminum sheeting, a box of rivets and the blueprints, might well be able to produce a serviceable wing, but would he or she deduce the principles of aerodynamic flight from copying the wing? By the same token, Mendel's laws could not have been predicted from the structure of DNA, or even that of chromosomes; in fact, they have meaning only in the context of cells and meiosis. It is common experience that to understand the whole we must know its parts, but the properties of the whole can seldom be predicted from the properties of its parts alone. That is what is meant by the chestnut that the whole is greater than the sum of its parts. And so, I find it unbelievable that the forms and functions of cells (let alone those of Weaver's primrose) will ever be predictable from a knowledge of their molecular constitution alone, however comprehensive. It would be a gross mistake to brush off the higher levels of biological order as if they were secondary or derivative;

on the contrary, how the parts come together must be key to any inquiry into the nature of life.

Schrödinger sensed this, and devoted the final pages of his slender book to the problem of biological order. The exquisite organization of every cell and organism appears to contravene the second law of thermodynamics, which insists that the universal tendency of physical processes is the dissipation of order and the production of entropy, a measure of disorder. Schrödinger credited the extraordinary ability of living things to generate, maintain and reproduce their orderly state to the extraction of "negentropy" (negative entropy) from the environment. Today, following Harold Morowitz (6), we would put this rather differently by saying that living organisms extract energy from the environment, use it to perform all manner of chemical and physical work, and thus convert energy into organization. Life does not contravene the second law; it evades it. But the problem remains that entities capable of converting energy into organization are not predictable from the laws established by classical physics. This suggested to Schrödinger that organisms stand outside physics in some essential respect; or else, that physics contains additional principles that pertain to organized systems, which remain to be discovered.

Can we discern any higher-order principles that are required for a fundamental understanding of life? One, at least, leaps to mind: Darwin's principle of evolution by random variation and natural selection. It has shaped molecules as much as organisms, and there is no explaining life without it. To be sure, the mechanisms that underlie evolution, like those of heredity or energetics, operate at the level of molecules, and some molecular processes that involve random variation and selection among macromolecules mimic biological evolution. But I doubt that evolution by natural selection would have been inferred from molecular science, had Darwin never lived; here is another generalization that finds full meaning only in the context of organisms. There may be others, such as the speculative proposition that the origins of biological form should be sought in the spontaneous self-organization of physical systems subjected to a flux of energy. Schrödinger, for one, considered that "living matter, while not eluding the 'laws of physics' as established up to date, is likely to involve 'other laws of physics' hitherto unknown, which, however, once they have been revealed, will form just as integral a part of this science as the former" (1). Here we touch one of the grand themes of a future biology, to which we shall return more than once.

Order, complexity, organization, function: these deceptively familiar words point the way toward the high intellectual frontier of biological science. Explorers who would travel that wilderness must put their trust, not in molecular biology alone, but in physiology—the science of complex systems. We all know in our hearts that a cell is far more than an aggregate of individual molecules; it is an organized, structured, purposeful and evolved whole. Unfortunately, analytical practice dictates that we begin our inquiries by grinding the exquisite architecture of the living cell into a pulp. No wonder, then, that the integrative perspective is woefully absent from the molecular view of life as it has developed over the past half-century.

So, what is life? The question is as good as ever. Despite decades of spectacular advances, the essential nature of life continues to elude us. We know much and explain more, but one sometimes suspects that our capacity to explain has outstripped our understanding. And when we have reinvented physiology, mastered self-organization and ransacked the rocks for fossil vestiges of genesis—will we then have read Schrödinger's riddle? Probably not. But we should have a much better grasp on the essential principles of the science of life, that grammar of biology for which Erwin Chargaff once wrote a memorable preface (5). At the very least, we should see more clearly what the riddle means, and how best to ponder it.

2

THE QUALITY OF LIFE

"The man behind the microscope
Has this advice for you:
Never ask what something Is
Just ask, what does it Do?"

Hilaire Belloc

We picked our way gingerly down the boulder-strewn canyon, keeping an eye out for cactus spines and rattlesnakes. One walks warily in the Big Bend, and so it was some time before we spotted our first "living rock." The nickname is apt: flat, grey and crusty, half-covered with sand, they blend into their stony environment. But once noticed, their nature is not in doubt; they are unmistakably living plants masquerading as rocks. It is almost always so: though the definition of life is elusive, we seldom have difficulty distinguishing living creatures from lifeless objects by their special qualities.

As a subject for serious inquiry, the category "life" has all but vanished from the scientific literature; it is the particulars of life, not its nature, that fill the numberless pages of scientific journals. But any attempt to extract general principles from that tide of information must begin, if not with a definition of life, then at least with the criteria by which we recognize the phenomenon. With the advent of space travel, the question has ceased to be purely an academic one. When explorers from

Starship Enterprise boldly land upon some planet orbiting Betelgeuze, will they recognize life if they encounter it in an unfamiliar guise? Probably yes, for wherever the restless search for novelty takes us, we expect life to be a quality of the peculiar class of objects called "organisms." They are the devil to define, but it is not difficult to set forth general criteria that map out the process of living as we see it all about us, and that should apply as well to life as we can imagine it elsewhere in the universe. Here are the main ones.

(i) *The Flux of Matter and Energy*. Living organisms are the seat of incessant chemical activity. They absorb nutrients, produce biomass and eliminate waste products plus heat; most constituents undergo breakdown and resynthesis during the lifetime of an individual organism. Metabolism, a term derived from the Greek word for change, designates the totality of all the chemical transformations carried out by an organism. It is so practical a hallmark of life that evidence of metabolism is what the space probe sent to Mars in 1976 searched for, without success.

Much of this chemical business revolves around energy. The characteristic activities of living things—their growth, movements, the very maintenance of their structure and integrity—depend upon input of energy from the environment. That is one of the chief functions of the metabolic web, for chemical substances serve as carriers of energy as well as matter. Like a flame or an eddy, an organism is not an object so much as a process, sustained by the continuous passage through it of both matter and energy.

(ii) *Self-Reproduction*. Living things are generated autonomously, not by external forces, and what they generate is their own kind. Like begets like. Biological heredity is quite unlike the point-by-point transfer carried out by a copying machine. Instead, characteristics are transmitted from parent to offspring by a program or recipe that embodies instructions for producing the next generation. The process is extremely accurate, yet subject to occasional errors that account for the variation observed in every natural population.

(iii) *Organization*. Whenever we speak of organisms we acknowledge the fundamental connection between the living state and a special kind of order. Even the simplest unicellular creatures display levels of regularity and complexity that exceed by orders of magnitude anything found in the mineral realm. A bacterial cell consists of more than three hundred million molecules (not counting water), several thousand different kinds of molecules, and requires some 2,000 genes for its specification. There is nothing random about this assemblage, which reproduces itself with constant composition and form generation after generation. A cell constitutes a unitary whole, a unit of life, in another and deeper sense: like the legs and leaves of higher organisms, its molecular constituents have functions. Whether they function individually, as most enzymes do, or as components of a larger subassembly such as a ribosome, molecules are parts of an integrated system, and in that capacity can be said to serve the activities of the cell as a whole. As with any hierarchical system, each constituent is at once an entity in itself and a part of the larger design; to appreciate its nature, one must examine it from both perspectives. Organization, John von Neumann once said, has purpose; order does not (1). Living things clearly have at least one purpose, to perpetuate their own kind. Therefore, organization is the word that sums up the essence of biological order.

(iv) *Adaptation*. Any organism that is made up of distinct parts, and that reproduces by heredity with variation, must evolve parts that promote the organism's survival and multiplication. Their structure and function will alter over time, tracking changes in both the internal and the external environment. The reason is that an individual's reproductive success must be affected by environmental factors, and natural selection will favor the better adapted over the less well adapted. Adaptive evolution is seen throughout the living world, not only at the level of legs and leaves but also at that of enzyme proteins and cellular organelles. That adaptation stems from the interplay of random variation and natural selection was, of course, Darwin's central contention. By recognizing adaptation as a criterion of life we do justice to life's intrinsic diversity. And we assert that the chemical and physical features of organisms find their meaning, first in the context of organization and then of history. Physiology and evolution are both central to the grammar of life.

With the help of these criteria we can quickly dispose of some doubtful cases. Is a flame alive? No. True, one candle lights another, but the size and shape of the flame are wholly determined by the supply of fuel and air, not by whether it was started with another candle or with a match. Fire propagates, but not by heredity. Viruses make a more interesting issue. They do propagate their kind by means of heredity, and they evolve and adapt all too quickly to changing circumstances; those who regard reproduction and adaptation as the crucial features of life will consider viruses to be alive. But viruses are structurally far simpler than cells, even than many organelles, they lack metabolism of any kind and are obligatory intracellular parasites. Their capacities are so much more limited than those of any cell that I, for one, would disqualify viruses. Much the same argument applies to mitochondria, and intra-cellular organelles in general: since the genes required for their production are located chiefly in the cell nucleus, organelles do not reproduce autonomously and must therefore be excluded from the ranks of the living. And what about freeze-dried bacteria? They were alive once, and provided they are "viable," may be alive again, but they are not alive at present. Such borderline cases are instructive rather than alarming. If life originated from the mineral realm via natural processes, we should expect the line that divides the quick from the dead to be a little fuzzy. Sharp categories are generally something that we put into nature, not something we find there.

It was the ambiguous status of viruses, whose crystallization had just been accomplished, that led N. W. Pirie to conclude that the terms "life" and "living" are inherently meaningless. That has not deterred his successors from proposing definitions, the best of which slip a kernel of truth into the nutshell of epigram (2). To J. Perret, "Life is a [property of] potentially self-replicating open systems of interlinked organic reactions, catalyzed stepwise and almost isothermally by complex and specific organic catalysts which are themselves produced by the system." Gail Fleischaker and Lynn Margulis, following the original proposal of Francisco Varela, make the point more succinctly and with sharper emphasis on the deep organizational features, when they define living organisms as "autopoietic systems," i.e., self-generating. Freeman Dyson puts himself in the same camp with the assertion that "life resides in organization, not in substance." Others are content with contemporary fashion; for Dulbecco, "Life is the actuation of the instructions encoded in the genes." Maynard Smith, however, points in quite another direction when he suggests that life might simply be defined "by the possession of those properties which are needed to ensure evolution by natural selection. That is, entities with the properties of multiplication, variation and heredity are alive, and entities lacking one or more of these properties are not" (2).

I have come to suspect that the definition of life is a mirror in which the various biological specialties chiefly see themselves. Functional biologists—biochemists, molecular biologists and physiologists—tend to look upon organisms as complex, integrated, and self-reproducing systems maintained by the stream of matter and energy. They ask how these systems work, and search for the proximal causes of the phenomena they observe in terms of physical and chemical mechanisms. Evolutionary biologists, by contrast, take a longer view. They ask how these systems came about and how their parts became mutually adapted.

Their hope is to discover ultimate causes, such as selective advantage or historical contingency, that shaped the patterns of form and function which we observe in all organisms. The secret of life is that these are two aspects of a single reality which we must strive to see in the round. No biological phenomenon can be said to be understood until we have found both its functional and its evolutionary explanation—and each of these is sure to be multilayered. To thread the maze of arguments woven about the relationship between living and non-living states of matter we must walk on two legs, one functional and the other evolutionary.

Of all the inanimate objects in the universe, few have so captivated the imagination of biologists as our own machines and automata. Nowadays it is the computer that is held up as the most instructive analog of living organisms, with cellular architecture as hardware and the DNA tape as software. Automata have complexity, functional parts and purposeful behavior just as living organisms do, but since they are man-made they carry no metaphysical baggage. Ever since Descartes there have been mechanistic biologists who see it as their task to "reduce" biology to chemistry and physics, for instance, to demonstrate that all biological phenomena can be completely explained in terms of the motions of their constituent parts and the forces between them. Biochemists and molecular biologists, in particular, commonly believe that such reduction is their objective, though they will not all agree on the meaning of the term. Some are satisfied that reduction has effectively been accomplished, thanks to the near-universal consensus that all that living things do is based on their physical substance, and that no metaphysical agencies or vital forces need be invoked. Many more would agree with Francis Crick (3) that "the ultimate aim of the modern movement in biology is in fact to explain all of biology in terms of physics and chemistry." And a few reductionists go still farther, maintaining that the laws and theories formulated in biology should be rephrased as special cases of those propounded in the physical sciences. That the two latter goals are illusory has been amply documented by George Gaylord Simpson, Michael Polanyi, Ernst Mayr and Alexander Rosenberg (4). Indeed, even a machine is not explained by mechanical principles alone, for its construction is guided by the designer's purposes which constrain the blind operation of physical laws. In the case of living organisms, it is their hierarchical organization and their origin in the interplay of random variation and natural selection that should give pause to any radical reductionist. And it is noteworthy that our unquestioned success in unraveling the molecular mechanics of life have thus far yielded little insight into the genesis of coherent forms and functions on the scale of cells and organisms.

For that reason, a majority of organismic biologists would probably be found aligned with an alternative general position, commonly known as holism (some prefer the more precise but awkward term "organicism"). Adherents hold that living organisms make up a set of unique, hierarchically organized systems each of which functions as a whole. Whenever a system is assembled from its constituent parts, novel properties emerge that could not have been predicted from a knowledge of those parts alone. The airplane wing that we contemplated in Chapter 1 is a case in point, and the argument applies *a fortiori* to any organism. Morphology, behavior and development are examples of such emergent properties that would never be inferred from molecular mechanisms, even if these were known in every particular. It follows that biology is an autonomous science (5), governed by laws and theories that emerge successively at the level of a cell, a frog, a flock of birds and a prairie pond. We can set aside, for the present at least, the question whether biology is autonomous in principle or only in practice, but we must note that holists feel the pinch of a shrinking domain. Time was when heredity and energy conversion were thought to be strictly the prerogative of living systems. Is it not likely that, given a few more decades, development and morphogenesis too will have been successfully reduced to the play of mindless molecules obeying only local rules?

I do not think so, and am often reminded of the arid quarrels over the nature of the Trinity that kept Byzantium in turmoil for centuries. Why should we be compelled to swear fealty to either reductionism or holism? Like John Tyler Bonner (6), "What is utterly baffling to me is why one cannot be a reductionist and a holist at the same time." Reductionism is commonly the best strategy in research, and when successful, supplies satisfying (albeit partial) explanations. Holists remember the inherent complexity of living things, and keep the reductionists honest. I was pleased to see that Hunter (6), re-examining the question whether biology can be reduced to chemistry, likewise takes a conciliatory position. The two extremes are complementary, not antagonistic: those who seek to understand living organisms require both the holist's perspective from the top down and the reductionist's scrutiny from the bottom up. Neither is sufficient by itself.

Many years ago, in a delightful essay celebrating the origins of molecular biology, Gunther Stent (6) spoke of the paradoxical quality of living things, which obey all the laws of physics and chemistry yet are not fully explained in terms of those sciences. Niels Bohr, Max Delbrück and Stent himself hoped to discover new laws of physics, hitherto unknown, that would supply physical and chemical explanations for the functions peculiar to life. No such laws have turned up, but one wonders whether we have been looking in the wrong direction. Biological phenomena of any interest are almost always properties of a system, more or less hierarchically organized into multiple layers. Simplification ("reduction") is commonly useful, even essential, to make a problem tractable, but it carries the risk of changing the question rather than answering it. To my mind, the beginning of wisdom is to recognize that living things are wholly composed of molecules, and everything they do finds a mechanistic explanation in terms of the actions and interactions of their constituent molecules. But their organization into systems of mounting complexity guarantees the emergence of supramolecular structures and activities. The more advanced the level of organization, the less informative is it to seek understanding solely in terms of their molecular constituents. It makes little sense to seek the molecular basis of hibernation because that is inherently the function of an organism (though one may hope to find genes and proteins specifically involved in hibernation). By the same token, the chemistry of leather is of little use in describing a shoe. Common sense suggests that we steer cautiously between molecular machismo and a veiled vitalism, some insights can be usefully expressed in molecular terms, others call for physiological explanations or for ideas appropriate to still higher levels of organization. We should be especially on the lookout for organizational principles that link molecules into cells and organisms, and for the historical forces that shaped the outcome. Common sense concurs with Paul Weiss that, "There is no phenomenon in a living system that is not molecular; but there is none that is wholly molecular either." For all their ubiquity and familiarity, living organisms are truly strange objects.

3

CELLS IN NATURE AND IN THEORY

"The cell is the microcosm of life for in its origin, nature and continuity resides the entire problem of biology."

W. S. Beck (1)

THE CELL THEORY
FROM FIVE KINGDOMS TO THREE DOMAINS
THREE PROFILES
REDISCOVERING THE ORGANISM

All through the eighteenth century microscopes improved in magnification, resolving power and optical clarity. In the middle of the nineteenth century, a large volume of observations on the tissues of higher plants and animals coalesced into a grand unifying conception, the "cell theory," rightly acclaimed in every textbook as a cornerstone of biological science. It states that all living things, notwithstanding their exuberant diversity, share a common architectural plan: every organism is composed of cells, either many or a single one, that constitute the fundamental units of life. It is a statement, not about the molecules of life and their chemical interactions, but about the spatial patterns into which these molecules are organized. In the hierarchy of biological order cells hold a special place, for they alone have the capacity to make themselves autonomously, and to multiply by division. Consequently, the cell represents the simplest level of organization that manifests all the features of the phenomenon of life. In the present chapter we shall examine how this conception arose, and how its meaning has evolved over the past century. For this purpose, our proper study is the world of microorganisms, whose manifold forms and lifestyles display the full range of options available to life in its most elementary mode.

THE CELL THEORY

The first to glimpse the throng of microscopic creatures, more than three hundred years ago, was Anton van Leeuwenhoek, merchant and civic official of Delft in the Netherlands, and an amateur lens grinder of consummate skill. His microscope, in effect a powerful magnifying glass, enlarged objects held on the point of a blunt pin by as much as 300 fold. It allowed van Leeuwenhoek to see protozoa and algae, cells of blood and tissues, sperm, even some bacteria; and he described this new universe in a barrage of enthusiastic letters (in Dutch) addressed to the Royal Society of London. Some of his specimens from 1674 have been preserved, and can still be identified (2). During the next two centuries, microscopists amply confirmed the abundance and diversity of microbial life. In fact, there are more kinds of life to be seen under the microscope than in the rain forests of the Amazon: of the major taxa, we now know, the majority are microbial.

Not until the mid-nineteenth century did microscopists begin to discern unity beneath the diversity of living forms. Cells had been observed many times since Robert Hooke coined the term in 1665 to describe the structure of cork, and as the limitations of early microscopes were progressively overcome, more than one biologist sensed that cells were universals rather than particulars. By mid-century the time was ripe for a unifying theory, which was forcefully articulated by the botanist Matthias Schleiden and the physiologist Theodor Schwann. Comparing notes over dinner one day in 1838, they recognized that the tissues of plants and animals were organized according to a common principle. Each consists of discrete droplets of a kind of jelly, soon to be named protoplasm, bounded by a membrane and enclosing a denser central mass or nucleus (Fig. 3.1). In separate publications that appeared in 1838 and 1839, Schleiden and Schwann presented their "cell doctrine:" plants and animals are not indivisible wholes but composites, made of innumerable cells, and each cell is itself an organism, endowed with the essential attributes of life. "Each cell leads a double life: an independent one, pertaining to its own development alone; and another incidental, in so far as it has become an integral part of a plant. It is, however, easy to perceive that the vital process of the individual cells must form the very first, absolutely indispensable fundamental basis, both as regards vegetable physiology and comparative physiology in general" (3).

Schleiden and Schwann held to the view that living matter arises by the aggregation or crystallization of non-living substance around the

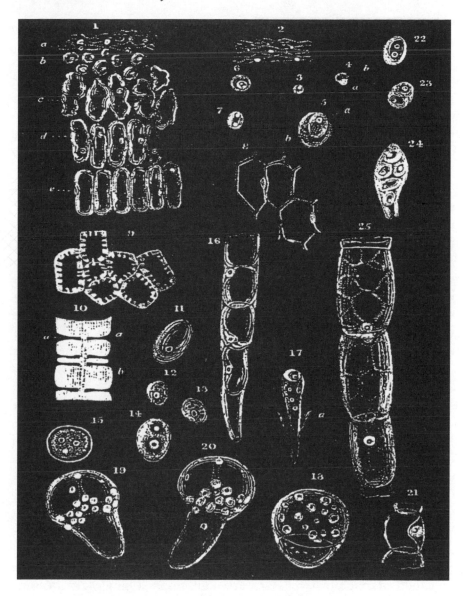

Fig. 3.1. Plant cells, as illustrated by Theodor Schwann in 1847. These include embryonic tissue (1), pollen tubes (16, 17) and germinating "sporules" (18–20). Photograph courtesy of the Wellcome Trust Medical Photographic Library, London.

nucleus. This quaint belief, already falling out of favor in their day, was soon abandoned; and Rudolf Virchow's textbook of pathology (1858) firmly asserted that cells, in fact, never arise spontaneously, but come always from pre-existing cells. A single cell may divide in two, or two gametes may fuse to make one, but cellular organization is continuous in either case. Virchow's categorical statement, *Omnis cellula e cellula* (Every cell from a cell) remains one of biology's essential verities (4).

The cell theory, originally formulated with higher plants and animals in mind, was extended to the microbial realm with the recognition (in 1845) that protozoa share the hallmarks of cellular organization: a boundary membrane and a nucleus. This discovery reinforced the growing conviction that cells are truly fundamental entities, the universal units of life. The growing interest in microscopic creatures had another consequence, whose bearing on the nature of living organisms only became clear in our own century. Ever since Aristotle, scholars and ordinary folk alike had divided the living world into two great kingdoms, plants and animals. Linnaeus, who devised the standard binomial nomenclature at the end of the eighteenth century, adhered to this view. Microorganisms were shoehorned into one kingdom or the other: the fungi and the sessile green algae into the plants, bacteria and the mobile predacious amoebae into the animals. But as information about microbial anatomy and physiology accumulated, it became clear that some combined the characters of both plants and animals—e.g., *Euglena*, which is green and photosynthetic yet vigorously motile. Others, such as the slime molds that are not partitioned into discrete cells, simply did not fit into either of the classical kingdoms. In 1866 Ernst Haeckel, Darwin's disciple and champion in Germany, redrew the tree of life with three great stems rather than two. He proclaimed a third kingdom, Protista, to accommodate the unicellular microorganisms and some of their multicellular relatives such as the marine and freshwater algae. The secession of the Protista involved far more than taxonomic convenience: it initiated a debate that continues today about the origin and early evolution of cells.

FROM FIVE KINGDOMS TO THREE DOMAINS

In unity, diversity again. Haeckel and his contemporaries regarded the multitude of microbial cells as variations upon a single architectural theme. But by the end of the century, with sharper and more powerful microscopes, it became apparent that bacterial cells were simpler and structurally quite unlike all other cells. The next major advance in our conception of what is meant by a cell was made in 1937 by the French cytologist E. Chatton, who distinguished two modes of cellular organization. He proposed the term *eukaryotic* (from the Greek for true nucleus) to designate cells that possess a discrete nucleus bounded by a special membrane, containing visible chromosomes during division. These are the cells that Schleiden, Schwann and their predecessors had in mind: not only the cells of higher plants and animals, but also those of fungi and many of the protists. They make up a diversified yet cohesive assemblage whose hallmarks include the true nucleus, intracellular organelles, a cytoskeleton, and an elaborate network of internal membranes; Bacteria, Chatton noted, were smaller, simpler and lacked a true nucleus, he proposed to call them *prokaryotic*. The significance of Chatton's contribution was largely lost on his contemporaries; it was left to R. Y. Stanier and C. B. van Niel, two decades later, to insist on the depth of the discontinuity that runs right across the biological world (5).

Taxonomists took note, however, for they had already seen need to revise the uppermost levels of the system of classification. It made no sense to lump together in a single kingdom organisms of radically different structure, and in 1938 bacteria were removed from the Protista into a kingdom of their own. What is today the standard classification was proposed by R. H. Whitaker in 1959, and is celebrated in the charming atlas of the living world prepared by L. Margulis and K. V. Schwartz (6). All living organisms are classified into five kingdoms, one prokaryotic (Monera) and four eukaryotic (Animalia, Plantae, Fungi and Protista); note the absence of viruses, which are not made of cells. All the bacteria, and only the bacteria, are prokaryotic in organization. Kingdom Monera therefore embraces the organisms called blue-green algae, familiar to everyone as the greenish scum that forms on the surface of standing water, and traditionally claimed by botanists as part of their turf. With the advent of the electron microscope these organisms were recognized as prokaryotes and rechristened cyanobacteria. Of the eukaryotic kingdoms the most interesting for present purposes is the Protista (7), a heterogenous collection of lower eukaryotes, most of them unicellular and but distantly related to one another. Slime molds, oömycetes, ciliates and flagellates are not household names, and most of them make little impact on our daily lives (except when one causes disease, such as *Giardia*). Yet among the protists we can expect to find, not only the ancestors of today's higher plants and animals, but the surviving descendants of the ancestral eukaryotic cells.

The doctrine that there exist two modes of biological organization, eukaryotic cells and prokaryotic ones, has become a cornerstone of our science and is firmly enshrined in every textbook. We are still coming to terms with the need to revise this familiar framework in light of the discovery that there are in fact three kinds of cells, one eukaryotic and two prokaryotic. Chatton and his successors defined eukaryotic organization by positive criteria, such as the possession of a true nucleus and of intracellular membranes. All that we have learned subsequently confirms the implication that eukaryotes, from protists to man, represent variations on a common set of biochemical, physiological and ultrastructural themes, and that they are united by descent from a common ancestor. By contrast, the prokaryotic pattern of organization was defined only by the absence of the distinctive eukaryotic features (5). The discovery that the term prokaryotes lumped together two classes of organisms that share the elementary ultrastructure, but differ profoundly in most other respects, is the most startling result of the application of molecular methods to the study of phylogeny.

Evolutionary relationships have traditionally been inferred from anatomical and physiological similarity: cats are obviously more closely related to tigers than to horses. This procedure is not suited to the bacteria, whose simple forms and functions offer few handholds. But in the sixties it was realized that the sequence of amino acids in proteins, or of nucleotides in nucleic acids (Chapter 4), contains a vast lode of genealogical information. Macromolecules that are descended from a common ancestral molecule diverge progressively over time, thanks to the accumulation of mutations; sequence comparisons can therefore provide insight into family relationships. Carl Woese recognized that, for microbiologists' purposes, the macromolecule of choice is the ribonucleic acid of ribosomes, cellular organelles that synthesize proteins (Chapter 4); the structure of ribosomal RNA is so stringently conserved that RNAs extracted from organisms that diverged billions of years ago still display unmistakable sequence similarity (8). Comparison of samples from a range of bacteria soon showed that, as judged by ribosomal RNA base sequences, prokaryotes fall into two entirely distinct groupings. The first, designated the eubacteria, includes most of the organisms familiar to students and to the general public: *Escherichia coli*, streptococci and pneumococci, cyanobacteria and other bacteria that degrade organic matter, make cheese and cause diseases of animals and plants. The second class consists of much more exotic organisms that tend to inhabit extreme environments: thermophilic bacteria, acidophiles, halophiles, rumen bacteria. These were designated archaebacteria because they were believed at the time to constitute a more ancient grouping. Bacteriologists had long been aware of the peculiarities of these organisms, regarding them as adaptations to their stressful environment. Woese and his colleagues showed, to the contrary, that the differences in RNA structure and in many other features represent an ancient divergence: eubacteria and archaebacteria are as distant from one another as each is from the eukaryotes (8).

During the past decade, as macromolecule sequencing became a routine operation, ribosomal RNA sequences emerged as the evolutionary biologist's chief quantitative tool—a clock to track the aeons. The universal phylogenetic tree shown in Fig. 3.2 summarizes the first fruits of this approach. The convention is that the length of each branch is proportional to the extent of divergence, measured not in years elapsed but in the number of mutations that resulted in an alteration of the nucleotide sequence in ribosomal RNA. The evolutionary distance between any two organisms is proportional to the length of the line that connects them. The chief conclusion leaps to the eye: all living organisms fall into three great clusters, for which Woese, Kandler and Wheelis (8) later proposed the term *Domains*: Eukarya, Eubacteria and Archaea. These taxonomic units correspond to, and supersede, the more familiar groupings eukaryotes, eubacteria and archaebacteria (9). We shall see in Chapter 8 that the tree has been rooted (albeit tentatively), and that Archaea and Eukarya emerge as distant, but specific, relatives. Domains are taxonomic units higher than the traditional kingdoms. As measured by ribosomal RNA sequences, the familiar kingdoms of animals, plants, and fungi are but twigs on the great bush of the Eukarya, with many

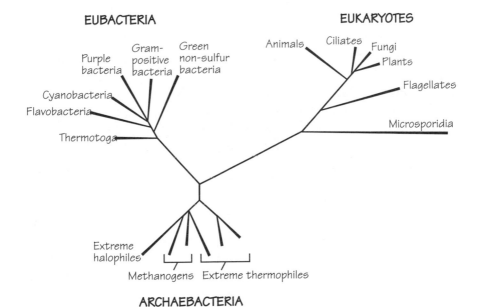

Fig. 3.2. Universal phylogenetic tree determined by comparison of ribosomal RNA sequences. Line lengths on the tree are proportional to calculated evolutionary distances. From Woese, 1987, with permission of the American Society for Microbiology.

additional branches corresponding to protists of one sort or another. Humans are much more closely related to mushrooms than to the bacteria that live on us and within us. Evidently, if animals and plants are to retain their traditional status as kingdoms, the total number of kingdoms will rise to near 30 among the Eukarya alone.

So drastic a restructuring of the global system of classification could not go unchallenged. The most perceptive critique came from Ernst Mayr (10), the dean of American evolutionists, who faults Woese and his colleagues on the very principles and purposes of their scheme. The universal tree with its three domains (Fig. 3.2) grows entirely out of differences in the structure of ribosomal RNAs, but variations in a single character make a notoriously fragile foundation for deep taxonomic divisions. The general organization of cells would provide a more robust platform, and from this viewpoint the salient fact is the prokaryotic nature of both eubacteria and archaebacteria. The differences that distinguish them, however significant, are apparent only to specialists in molecular science. Eukaryotes, by contrast, represent an enormous evolutionary advance, "surely the most drastic change in the history of the organic world," which produced altogether novel and more complex cells. Mayr, therefore, advocated that global classification be squarely based on the level of cellular organization: two domains, one prokaryotic (with eubacteria and archaebacteria as subdomains) and the other eukaryotic. In principle, though not in every detail, this classification appeals to many students of microbial life (10), and I also find much merit in it.

In a forceful rejoinder to their critics, Wheelis et al (11) point out that a nucleotide sequence is not to be taken as a single character: as a long string of units that can vary independently, a molecular sequence is far richer in genealogical information than the classical characters of anatomy and function. It is also closer to the locus of variation: "[It] is at the level of molecules (particularly molecular sequences) that one really becomes privy to the workings of the evolutionary process. Molecular sequences can reveal evolutionary relationships in a way and to an extent that classical phenotypic criteria, and even molecular functions, cannot; and what is seen only dimly, if at all, at higher levels of organization can be seen clearly at the level of molecular structure and sequences. Thus, systematics in the future will be based primarily upon the sequences, structure, and relationships of molecules, the classical gross properties of cells and organisms being used largely to confirm and embellish these." One need not swallow the rhetoric whole to concur that Mayr's two-domain scheme puts tradition and convenience ahead of the data, whose import is quite plain. Granted the premise that ribosomal RNA sequences track the evolution, not just of ribosomes but of organisms, it follows that a "natural" classification, one that mirrors the lines of descent, must accommodate three primary clusters.

For the purposes of this book we need not take sides in the debate over the proper rank to be assigned to each cluster, but we do need to recognize that, as matters presently stand, there are three of them. This seems to me of such importance as to override other considerations, and therefore the remainder of this text will assume a tripartite division of the living world. The term "prokaryote" remains useful, but it describes a grade of cellular organization rather than a population of related organisms. The number three has carried overtones of sanctity, but in the present instance it is merely provisional, for it is entirely possible that additional domains will be discovered as biologists probe remote and unfamiliar environments. Who is to say what kinds of organisms inhabit the clefts and fissures of the Mid-Atlantic Ridge, miles beneath the ocean floor?

THREE PROFILES

In their classic article on the concept of a bacterium, Stanier and van Niel (5) concluded that "the distinctive property of the bacteria and blue-green algae is the procaryotic [*sic*] nature of their cells." They listed a series of criteria that distinguish bacteria from eukaryotic protists, underscoring the simplicity of form and ultrastructure of the prokaryotes, but had no inkling that this grade of cellular organization describes both Eubacteria and Archaea (Fig. 3.3). Prokaryotes possess but a single chromosome, or genophore, circular as a rule. Cells are enclosed in a plasma membrane but lack a nuclear membrane and (with some exceptions) endomembranes in general. Energy production is usually a function of the cell as a whole, not of specialized organelles. The cytoplasm is almost featureless aside from the numerous ribosomes and occasional inclusions; the former are smaller than eukaryotic ribosomes and sensitive to a distinct set of antibiotics. Prokaryotes do not display a permanent cytoskeleton, and specifically lack both microtubules (tubulin) and microfilaments (actin). The complex motility organelles of eukaryotic cells are never found. Instead, prokaryotes swim with the aid of rotating flagella; each flagellum is made up of monomers of a protein, arrayed into a helical filament that is plugged through the plasma membrane. Most prokaryotic cells are walled. The characteristic component of the wall is a peptidoglycan, a co-polymer of sugars and amino

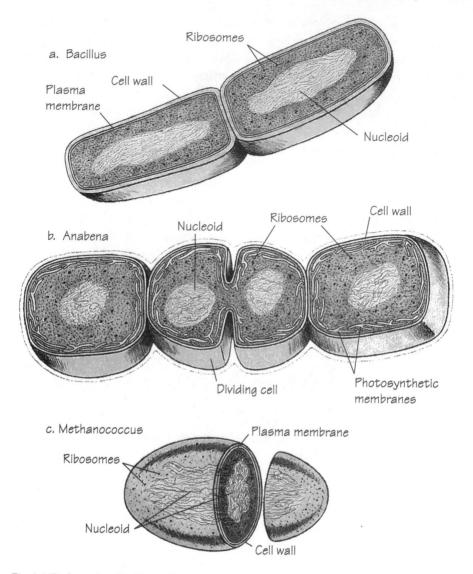

Fig. 3.3. Prokaryotic cells. (a) *Bacillus*, a typical eubacterium with minimal cytoplasmic structure. (b) The cyanobacterium *Anabena* with prominent cytoplasmic membranes. (c) *Methanococcus*, a thermophilic archaebacterium from a deepsea vent. Cell lengths, 1-5 μm.

acids, rather than the chitin and cellulose that feature in eukaryotic cell walls. Finally, prokaryotes never harbor endosymbionts. Their plain architecture stands in sharp contrast to their metabolic diversity: prokaryotes have evolved so as to exploit the full range of energy sources available on earth. We find among them aerobes and anaerobes, all the known kinds of photosynthesis including some unique sorts, the capacity to oxidize H_2, H_2S and Fe^{2+}, to reduce sulfate to sulfide and CO_2 to methane, and also the machinery for nitrogen fixation.

Fig. 3.3 illustrates representatives from both the Eubacteria and the Archaea in order to demonstrate that the features that lead one to recognize these as fundamentally different organisms are not apparent at the ultrastructural level; but at the molecular level the differences are extensive. Ribosomal RNAs differ between the two domains, not just in sequence but in the kinds of sequences they contain. Languages furnish an analogy. English, Italian, and Persian all belong to the Indo-European cluster; they share not only words but a grammar. Chinese is constructed on an altogether different basis, Arabic on yet another. Aside from RNA sequences, which were the original basis for distinguishing Eubacteria from Archaea, we now have additional criteria to buttress the distinction. Cell walls of Eubacteria always contain a particular kind of peptidoglycan, commonly designated murein. Cell walls of Archaea lack murein; some contain a related peptidoglycan, others are made of protein. The plasma membranes of Eubacteria are made of phospholipids, those of Archaea contain branched ether lipids instead. Ribosomes of Archaea lack certain proteins that are found in the ribosomes of Eubacteria, and RNA polymerases from Archaea have many more subunits than do those of Eubacteria. There are differences at the level of metabolic pathways, of coenzymes, of membrane transport proteins, all on display as more and more genomes are sequenced in their entirety (12). There is considerable confusion as well (Chapter 8), and it is not at all easy to identify criteria that distinguish all the Archaea from all the Eubacteria. Consistent differences have emerged in two areas, both of fundamental importance: membrane chemistry and the core mechanisms by which genetic information is expressed.

Readers who are not microbiologists or biochemists may balk at the claim that such purely molecular criteria justify the creation of a separate domain Archaea, equivalent in taxonomic rank to that which houses all the eukaryotes from *Paramecium* to the blue whale. But reservations concerning their proper taxonomic status must not blind one to the magnitude of the discovery. Since morphological and physiological characters are not useful guides to the early stages of cellular evolution, we must perforce rely on molecular criteria, and Archaea clearly represent a major divergence from some early ancestor of cellular life. It is also no longer possible to dismiss the Archaea as denizens of obscure and narrow niches that play but a minor role in the biosphere as a whole. The deeper portions of the earth's crust contain many potential habitats suited to microorganisms adapted to heat, high pressure, acids and inorganic sources of chemical energy (13); and hitherto unknown and unculturable Archaea make a major contribution to the microscopic plankton of the open oceans. Recent calculations (13) confirm what microbiologists have long suspected, that prokaryotes make up the "unseen majority" of life as judged by biomass as well as cell numbers; a goodly proportion of that majority probably consists of Archaea.

In contrast to the prokaryotic realms, whose taxonomic status is debatable, there is no doubt that the Eukarya represent a unitary assemblage, however deeply diversified. Chatton had stressed the possession of a true nucleus, complete with nuclear membrane and discrete chromosomes, as the diagnostic feature. With the advent of the electron microscope, it became clear that the differences between eukaryotes and prokaryotes are both wider and deeper (14): they represent two distinct ways of organizing a cell. Three eukaryotic cells, a unicellular protist, a yeast cell and a plant cell, are sketched in Fig. 3.4 in order to document that, however different they are in gross morphology and habits, they are all made from a common set of standard parts.

(i) Eukaryotic cells are much larger than prokaryotic ones, typically of the order of 10 micrometers or more in diameter. Since the volume increases as the cube of the radius, it follows that the volume of a eukaryotic cell is typically a thousand times larger than that of a prokaryotic one. This may be the most basic difference between the two. Surface area rises as the square of the radius, whereas the volume rises as the cube; consequently, larger cells have a lower ratio of surface to volume. It seems likely that the complex architecture of eukaryotic cells, especially their extensive system of endomembranes, represents an adaptation to their larger size. The argument reads better the other way around: internal membranes made larger cells workable (15).

(ii) Energy metabolism in eukaryotic cells is the business of specialized, discrete intracellular organelles. Mitochondria and chloroplasts are familiar examples. More recondite are hydrogenosomes, organelles found among anaerobic protists, which house enzymes of anaerobic metabolism.

(iii) Cytoplasmic traffic in macromolecules does not rely on diffusion, as in prokaryotes, but involves a ramified system of endomembranous organelles linked by mobile vesicles. The endoplasmic reticulum, Golgi apparatus, lysosomes, and vacuoles are parts of this system.

(iv) The organelles of movement found in eukaryotic cells are quite unlike those of prokaryotes. Undulipodia (a term that designates specifically the eukaryotic kind of flagella and cilia) are complex structures built of microtubules, arrayed in a characteristic pattern of nine periph-

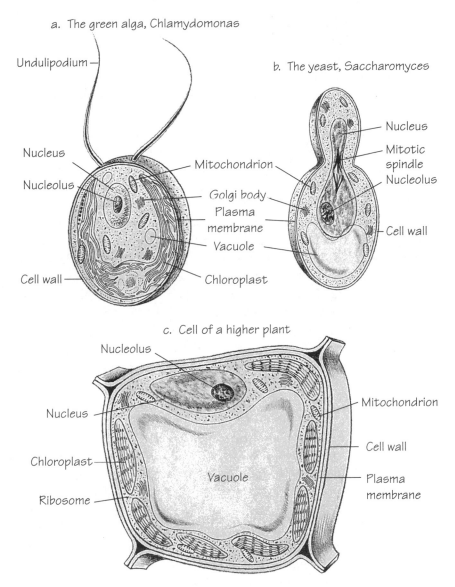

a. The green alga, Chlamydomonas

Undulipodium

b. The yeast, Saccharomyces

Nucleus

Nucleolus

Mitochondrion

Golgi body

Plasma membrane

Vacuole

Cell wall

Chloroplast

Nucleus

Mitotic spindle

Nucleolus

Cell wall

c. Cell of a higher plant

Nucleolus

Nucleus

Chloroplast

Ribosome

Vacuole

Mitochondrion

Cell wall

Plasma membrane

Fig. 3.4. Eukaryotic cells, highlighting common organelles. (a) The green alga *Chlamydomonas*. (b) Budding yeast cell, a fungus. (c) A higher-plant cell. Cell bodies, 10 μm (5 μm for yeast).

eral tubules and two central ones. Movement results from the coordinated sliding of one microtubule filament relative to its neighbor, mediated by a motor protein called dynein. Most (probably all) eukaryotic cells contain a second system of motility, based on microfilaments and myosin. Perhaps the essential difference is that eukaryotic cells contain mechanoenzymes that transduce metabolic energy (ATP hydrolysis;

Chapter 4) into mechanical work; such enzymes are absent in prokaryotes, or at least rare.

(v) In prokaryotes, mechanical integration of the cell is the business of the plasma membrane and cell wall. For instance, genophores separate at the time of cell division by being linked to the expanding cell wall. Eukaryotic cells possess an elaborate cytoskeleton, based on microtubules and microfilaments, that serves the functions of scaffolds, struts, tracks and pulleys. It is noteworthy that the molecular building blocks of the cytoskeleton, tubulin and actin, are absent from prokaryotes, albeit distant molecular relatives can be identified.

(vi) The eukaryotic genome is larger than the prokaryotic one by one or more orders of magnitude, and it codes for a correspondingly larger number of proteins. Eukaryotic genes are typically interrupted by noncoding sequences, and long stretches of apparently meaningless DNA are interspersed among the working genes; a pattern aptly described as islands of functional DNA in a sea of non-functional DNA. The genome is divided into discrete chromosomes built around proteins called histones; their precise distribution when the cell divides is the task of the mitotic apparatus. In addition, most, but not all, eukaryotic cells contain multiple genomes: mitochondria and chloroplasts house small genomes of their own, and further examples may yet be discovered.

REDISCOVERING THE ORGANISM

Scientific theories are commonly formulated with a purpose, in an effort "to explain visible events by invisible forces, to connect what is seen with what is assumed," as Francois Jacob once put it (16). Darwin's theory of evolution by natural selection is a renowned case in point, Peter Mitchell's chemiosmotic theory of energy coupling a less familiar one (Chapter 5). In this respect, the cell theory is rather an anomaly. To begin with, at least, a "cell" appeared to be no more than the abstract description of a recurrent cytological motif, devoid of explanatory power. But the idea wears its years well: today we acclaim it as the first recognition of the unity of all life on earth. Beneath the diversity of forms and specialized functions, cells share a common architecture that extends even to creatures such as slime molds, some of which have no cells in the usual sense. All are made up of a limited number of standard parts—ribosomes, chromosomes, membranes—arranged in endless permutations. Scientists began to perceive a new objective on the far horizon. "With the cell, biology discovered its atom. . . . To characterize life, it was henceforth essential to study the cell and analyze its structure: to single out the common denominators, necessary for the life of every cell; alternatively, to identify differences associated with the performance of special functions" (16).

During the past half-century, the program of analyzing the structure of cells in search of their common denominators has been pursued to the molecular level with notable success and with mounting zeal. This single-minded concentration on the relatively tractable problems of chemical structures and interactions has been accompanied by neglect of the higher levels of biological order, often to the point of absurdity. Surely, one plain lesson to be read in the prevalence of cellular architecture is that organized complexity is one of the essential characteristics of life. From this viewpoint, the significance of cells is that they represent the minimal level of organization capable of displaying the activities we associate with life, including self-reproduction. In a nutshell, cells are the simplest autopoietic systems in our kind of world. That was perfectly clear to J. H. Woodger, writing more than seventy years ago. He regarded it as "something of a scandal" that biologists have no adequate concept of organization: "The failure to take organization seriously is perhaps but another consequence of the rapid development of physics and chemistry as compared to other sciences, and the consequent dazzling effect this has had on biological vision" (17). The thesis that cells can only be understood as organized systems, and that the physiological and molecular levels of inquiry are necessarily complementary, is one of this book's recurrent themes. That this self-evident proposition should remain in practice so much of a minority view never ceases to astonish me.

It seems to be true of theories, as of people, that as we grow older the world becomes stranger, the pattern more complicated. What began as an illuminating vision of the cell as the atom of life, now looks more like the sphinx at the crossroads posing awkward riddles. Why, for example, should there be three kinds of cellular patterns? What selective pressures favored the emergence of ornate eukaryotic cells in some situations and of spartan prokaryotic ones in others? What is it that gives the archaeal pattern the edge in various extreme environments? Are all aspects of cellular architecture products of natural selection, or do some bespeak deeper laws of order that govern complex patterns? What is the nature of that universal ancestor that gave rise to all extant forms of life? And then there is that mystery of mysteries, how did cells arise before there were cells? Such questions point beyond both molecular science and physiology to a world of historical contingency; most of them we hardly know how to ask.

4

Molecular Logic

Marco Polo describes a bridge, stone by stone.
"But which is the stone that supports the arch?" Kublai Khan asks.
"This bridge is not supported by one stone or another," Marco Polo answers, "but by the line of the arch that they form."
Kublai Khan remains silent, reflecting. Then he adds, "Why do you speak to me of the stones? It is only the arch that matters to me."
Polo answers, "Without stones there is no arch."

Italo Calvino (1)

A Protein for Every Task
Membranes: Keeping Apart, Bringing Together
Energy, Work and Vitality
Genetic Information
Protein Folding: Giving Meaning to Message
Keeping Control: Regulation and Homeostasis
Self-Assembly: Molecules into Structures
The Unity of Biochemistry
Molecular Evolution: The Mark of the Tinkerer

When biochemists set out to tackle a problem, our first step is commonly to grind the intricate fabric of cells and tissues into a pulp (a homogenate, as we say in the trade). This is a significant act, representing a drastic reduction in the level of organization. It allows us to treat living matter as a mixture of chemicals, and encourages one to isolate and purify individual constituents; every graduate student is warned not to squander clean thinking on dirty enzymes! To be sure, something is sacrificed by this violent procedure—not only life itself, but all the spatial order that impresses anyone who inspects a photomicrograph.

But never mind: biochemists still cherish the premise that nothing ir-retrievable is lost by homogenization, and that given the macromole-cules, all the essentials are present and accounted for. We know quite well that this cannot be true, but the focus on the molecules defines that layer of knowledge that we designate as biochemistry or molecular biology, and undergirds our professional identity.

Let us for the present set aside the levels of order lost to the tissue grinder, and celebrate the astonishing achievements that grew from the meticulous examination of life's fragments. All the activities of living things are carried out by molecules, and are thus ultimately rooted in molecular structures and interactions. Without subscribing to the lop-sided view that, therefore, all of biology is molecular biology, it must still be said that one cannot reflect usefully on the phenomenon of life without taking account of its material basis. There can be no arch with-out stones. Besides, the molecular level provides an excellent introduc-tion to the exploration of biological order in general. Regularity, pur-pose, and complexity pervade the field. We can observe individual molecules coming together to perform novel and emergent functions, each of them a whole larger than the sum of its parts. We can discern generalizations that apply to all organisms on earth; they qualify locally as laws of biology, and assure us that all life is of one kind. We also find variations on all but the most basic themes; these report the roles of mutation and selection, constraints and contingency, in the genesis of living order. And we can ponder the deep question, what might we expect to find if and when we encounter life beyond the solar system.

Biochemists and molecular biologists revel in the details of their sub-ject; the key to finding order in the profusion is the concept of *function*. "Living organisms are composed of lifeless molecules," the late Albert Lehninger proclaims on the opening page of his classic textbook (2), but those molecules are special. The molecules of life differ from those encountered in the inorganic world, not in their chemical qualities, but in their biological ones: with a few exceptions, such as waste products, each performs a job in the service of the organism as a whole. The notion of function is meaningless when applied to the constituents of clay or of petroleum, for those molecules are the products of physical and chemical forces alone, but function becomes crucial when we ask why leaves are green and blood is red. Function implies purpose, and therefore, order. "The molecules of which living things are composed conform to all the familiar laws of chemistry, but they also interact with each other in accordance with another set of principles, which we shall refer to collectively as the *molecular logic of the living state*" (2). In this chapter, we will be walking in Lehninger's trail. Most of its factual content is drawn from standard textbooks, and students of biochemistry will find little in it that is new to them, but for the layman it may be like sipping from a firehose. I have sought to make this chapter useful to the general reader by underscoring the principles and omitting the details; some of the latter will be found in Boyce Rensberger's useful book (3), which is specifically addressed to a lay audience.

For the purposes of this chapter, we may think of a cell as an intricate and sophisticated chemical factory (Fig. 4.1). Matter, energy and information enter the cell from the environment, while waste products and heat are discharged. The object of the entire exercise is to replicate the chemical composition and organization of the original cell, making two cells grow where there was one before. Even in the simplest cells, this calls for the collaborative interactions of many thousands of molecules large and small, and requires hundreds of concurrent chemical reactions. These break down foodstuff, extract energy, manufacture precursors, assemble constituents, note and execute genetic instructions and keep all this frantic activity coordinated. The term "metabolism" designates the sum total of all these chemical processes, derived from the Greek

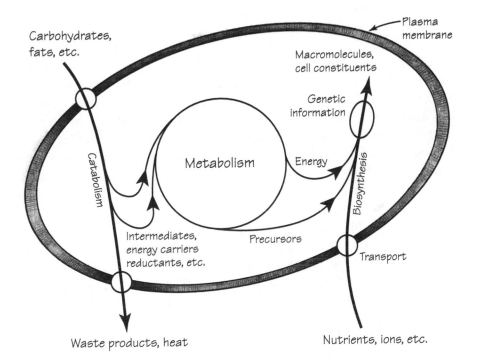

Fig. 4.1. The cell as a chemical factory.

word for "change." Biochemistry, then, is the study of the chemical basis of all biological activity. My purpose here is of necessity much more modest: to introduce the reader to terms, discoveries and ideas that are indispensable for closer reflection on the nature of life.

A PROTEIN FOR EVERY TASK

A beginning student, on first encountering the profusion of chemical reactions that take place in every cell, is apt to suspect that any reaction that can possibly go, does go. Closer inspection corrects the false impression: in actuality, cellular metabolism is highly selective and quite purposeful. Each reaction is mediated by a particular enzyme whose function is to enable that reaction to proceed at a high rate, often with extreme specificity and with minimal formation of useless by-products. Step by simple step, the cell's complement of enzymes breaks down foodstuffs, turns them into metabolites and then into cell constituents, and harnesses the energy of some reactions to drive others (Fig. 4.1). Enzymes select the channels through which matter and energy flow. They can be studied as single molecules and often are, but they derive meaning from being parts of a larger whole, the metabolic web.

How enzymes perform their catalytic feats, greater by many orders of magnitude than those of inorganic catalysts, has long been one of the central questions in biochemistry. The heart of the matter is the specific, intimate, and tight binding of the substrate (or substrates) to the enzyme. Proteins (and virtually all enzymes are proteins) are not shapeless blobs, but sculptured objects, equipped with crannies and cavities that admit particular molecules, while excluding others. Binding commonly entails changes in the configuration of both substrate and enzyme, inducing stresses and strains that contribute to the mechanism of catalysis. Besides, the catalytic site supplies chemically active groups in the form of amino acid side-chains that actually participate in the reaction. The catalytic site is tailored, as it were, to its particular task, linking its structure to its function.

The genome of *E. coli* encodes approximately 4,000 proteins, that of yeast 6,000; it takes 50,000 proteins or more to make a man. What do they all do? Many proteins are enzymes, but by no means all. Some proteins serve as the building blocks of structural scaffolding. Some make tracks for the movement of organelles, itself mediated by motor proteins. Proteins act as receptors for signals from within the cell or from the outer world; they transport nutrients, waste products and viruses across membranes. Proteins also commonly modulate the activities of other proteins, or of genes. The general principle is that, except for the storage and transmission of genetic information and the construction of compartments, almost all that cells do is done by proteins.

The explanation for the functional versatility of proteins is not chemical so much as physical. Amino acid chains can fold into a variety of shapes, globular and fibrous, each determined by the sequence of the amino acids that make up the protein in question. As they fold, each generates a unique contour with its own pattern of structural features: rods and hinges, platforms and channels, holes and crevices. Moreover, proteins are flexible and dynamic constructs that commonly change shape when they interact with ligands or with each other. The range of stable configurations that amino acid chains can assume is wider than that of other classes of macromolecules, nucleic acids in particular; and their flexibility permits all sorts of mechanical actions demanded of molecular machines.

Proteins, as catalysts and structural elements, are part of biochemical tradition; more recently we have come to see many of them as mechanical devices that rely on energized motion to perform their tasks. Even enzymes can be profitably looked at from this point of view: with the growing catalogue of enzyme structures has come the recognition that active sites and their elements commonly undergo rearrangement as part of the catalytic cycle and its regulation. Other proteins are there to bring about overt movement, either of molecules or of larger objects. Transport carriers reorient the binding site from one membrane surface to the other, and back again; sometimes the mechanical cycle is coupled to an energy source, turning the carrier into a pump. Students of eukaryotic cells are finding ever more motor proteins that translocate vesicles, chromosomes, or elements of the cytoskeleton from one place to another. The most familiar example is myosin, whose cyclic change of conformations underlies muscle contraction and some instances of cell motility. And bear in mind ribosomes and the polymerases that transcribe and replicate genetic information: energized movements are central to their operations. As we unravel the molecular workings of life, the cell presents itself as an assemblage of tiny machines; mundane mechanical engineering looms as large as the subtle flow of energy and information.

Few generalizations in biology get by without qualification, and that applies to the status of proteins as the tools for all tasks: surprisingly, some tasks fall to ribonnucleic acids. RNAs make up almost two thirds of the ribosome's mass, and play a catalytic role in linking up the amino acids during protein synthesis. They contribute to the structure and function of several less familiar organelles, such as the particles involved in the translocation of proteins across membranes. Most remarkably, RNA alone sometimes serves as a catalyst. Catalytic RNAs, or ribozymes, participate chiefly in the manipulation of RNA, and they do not challenge the predominance of proteins in the catalysis line. But their discovery in the eighties overthrew the universal consensus that all enzymes must be proteins, and it suggested an altogether novel line of inquiry to students of the origin of life.

Membranes: Keeping Apart, Bringing Together

The composition of living cells is grossly different from that of their environment. That is possible because each cell is enclosed by a membrane, extremely thin and flexible, that is essentially impermeable to molecules large and small. Biological membranes are composed of phospholipids (or, in Archaea, of the related ether-lipids), that form closed bilayered structures spontaneously (Fig. 4.2). The oily core of the bilayer excludes water-soluble molecules, and thus imposes a barrier to the diffusion of most substances; water itself, oxygen and other gases are among the exceptions. Membranes are as essential to life as genes and proteins: "To stay alive you have to be able to hold out against equilibrium, maintain imbalance, bank against entropy, and you can only transact this business with membranes in our kind of world" (4).

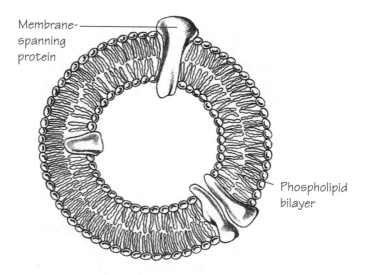

Membrane-spanning protein

Phospholipid bilayer

Fig. 4.2. The architecture of biological membranes: A phospholipid bilayer makes up the basic structure; proteins that span the membrane mediate the passage of nutrients and other molecules.

The corollary is that cells require special means to transport nutrients in and waste products out. With very few exceptions, this is accomplished with the aid of proteins that span the membrane, linking the aqueous phases inside and out. Transport proteins are akin to enzymes in recognizing and binding particular substrates, but most of them do not catalyze any chemical reaction, they merely translocate the substrate from one side to the other by shifting the orientation of the binding site from one surface to the other. In some cases this reorientation is linked to an energy source, allowing the transport protein to "pump" its substrate against a concentration gradient, others merely facilitate downhill diffusion. The plasma membrane of a bacterial cell may contain as many as a hundred different transport catalysts, each more or less specific for a particular substrate.

Transport proteins carry matter; receptors deal in information. Bacterial cells sense the presence of nutrients in the medium by virtue of specialized proteins that span the plasma membrane. When a potential nutrient binds to the receptor protein, the news is carried across the membrane by a conformational change; this, in turn, activates an enzyme or an enzyme cascade. What passes across the membrane is neither matter nor energy, but a signal: "now," or sometimes "here."

Bacteria, as a rule, feature but a single membrane, the plasma membrane that defines the cell; but in eukaryotic cells, intracellular membranes are a conspicuous feature. Each intracellular membrane encloses a defined space whose composition and function is more or less distinct, and which communicates with the cytoplasmic fluid by means of specialized transport catalysts. These compartments are also dynamic, fusing with one another and detaching in a controlled and functional manner.

Traffic across membranes is not confined to small nutrients and waste products. In a bacterial cell, many constituents are situated external to the plasma membrane: the cell wall, an array of enzymes, proteins that serve as receptors for environmental signals, flagella, and other surface organelles. The cells must thus secrete molecules of considerable complexity, including wall precursors and whole proteins, and they must meanwhile maintain the integrity of their bounding membrane. The role of membranes, like that of the sea in human history, is dual: they separate compartments while channeling the flow of matter and of information. When one thinks about membranes, one stands on the border between molecules and cells.

ENERGY, WORK AND VITALITY

The flux of matter through cells and organisms is a fact of daily experience; the flow of energy is imperceptible but just as fundamental, and both are mediated concurrently by that intricate web of chemical reactions that make up cellular metabolism. The ability to capture energy and to harness it to the performance of diverse kinds of work is a sine qua non of life. With the recognition of energy and its roles in biology, the search for the vital force reached its terminus.

The cellular economy can be abstractly represented by the flow diagram in which nutrients, energy and information are turned into more cells, waste products and heat (Fig. 4.1). Growth, reproduction and repair entail the manufacture of molecules large and small, each synthesized by a succession of chemical steps. Most of these steps can be described as "work" in the sense that they would not occur spontaneously. Take proteins. The natural tendency of proteins is to fall apart; for proteins to be synthesized, the reaction must be driven up the thermodynamic hill, away from equilibrium. The same is true of other biochemical processes: the transport of nutrients against a concentration gradient, the generation of physical force or electrical potentials, even the accurate transmittal of genetic information, all represent work in the thermodynamic sense. They can take place only because cells couple the work function to a source of energy. This, in fact, is how energy is defined: it is the capacity to do work. Bioenergetics revolves around the sources of biological energy, and the mechanisms by which energy is coupled to useful work (5).

The universe is ablaze with energy, but living organisms rely on just a few kinds. The ultimate energy source for the great majority of biological processes is sunlight. A small fraction of the light that reaches the earth's surface is absorbed by chlorophyll, a family of molecules found in the photosynthetic apparatus of algae, plants and certain bacteria. Light absorption initiates a complicated sequence of chemical reactions that convert (or transduce) the energy of light into chemical form (Fig. 4.3). The energy is then employed, again through chemical transactions, to perform the many tasks that living things find useful, and is finally degraded to heat and radiated into space. Organisms that cannot carry out photosynthesis (animals, fungi, and the majority of bacteria) obtain energy from the breakdown of organic substances, commonly by oxidation (respiration). Since organic matter was originally produced by photosynthesis, virtually the entire biosphere subsists on the sun's beneficence. The ancients were not mistaken when they worshiped the sun as the fount of life!

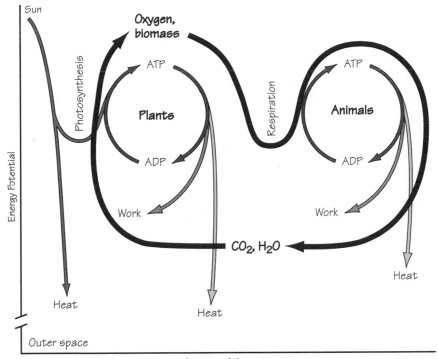

Fig. 4.3. Energy flow through living organisms. A fraction of the light energy that reaches the earth is captured by photosynthetic organisms, in the form of biomass and ATP. Animals and most microorganisms generate ATP by degrading organic matter, originally produced by photosynthesis. ATP is broken down during the performance of work. Ultimately, all the energy is dissipated as heat. The heavy line designates the flow of carbon, the lighter one that of energy.

The exceptions to this heliocentric view of global energy flow take the form of bacteria that make their living from inorganic processes, such as the reduction of carbon dioxide to methane or the oxidation of hydrogen sulfide. Some of these "chemolithotrophic" bacteria have been studied in detail, particularly the methanogenic ones that inhabit the cow's rumen, but the great majority are little known even to microbiologists. Many chemolithotrophs belong to the domain Archaea. They often flourish in unconventional niches that are anaerobic, highly acidic or very hot, which makes them difficult to study; many have never even been grown in pure culture. The standard practice has been to regard chemolithotrophs as curiosities, biochemically interesting but of minor significance to the earth's energy budget; fifteen years ago I gave them short shrift (5). But this may well be a false perspective, for two reasons. First, bacteria are increasingly turning up in places previously thought to be sterile, particularly the deep, hot rocks of the earth's crust. Habitats suitable for organisms that can draw on geochemical processes for energy now appear so extensive that their denizens may make up a large fraction of the world's total biomass (6). Second, there are reasons to suspect that the earliest organisms relied on inorganic energy sources (7). Should this prove true, it would transform our perception of both global energy flow and its relationship to the origin of life.

Energy is to biology what money is to economics, the means by which organisms purchase goods and services, and since most of the latter are chemical in nature (e.g., biosynthesis), energy must be supplied in chemical form. A number of small molecules can serve as energy donors for biochemical reactions, and one of these functions as the universal energy currency: adenosine triphosphate, ATP (Fig. 4.4a), which is a member of the core set of cellular metabolites. ATP participates chemically in most of the processes that it promotes. The adenosine group serves as a handle that binds to proteins that process ATP, the three phosphoryl groups are the working end of the molecule. The terminal phosphoryl group (sometimes the last two) is split off in the course of energy transfer to give the diphosphate (or the monophosphate, respectively). It is irresistible, albeit inaccurate, to envisage metabolic energy being conserved, or stored, in the "high-energy bonds" that connect the phosphoryl groups. To set the record straight, it is the marked tendency of ATP to give up its phosphoryl groups, either to water or to other molecules, that makes it a suitable vehicle for the transfer of chemical energy. For many purposes, we can assume that biological energetics consists of the processes that generate ATP (and other "energy-rich" metabolites), and those that consume these metabolites (Fig. 4.4b). As in economics, the linkage between supply and demand is intensely dynamic. The great highways of energetics, respiration and photosynthesis, keep the ATP/ADP ratio high, far away from equilibrium, and that, in turn, allows ATP to serve as an energy donor, displacing from equilibrium those reactions in which it participates. All biosynthetic processes, and also those that entail movement or transport, are energized either by ATP or by one of the more specialized energy carriers, and the latter are linked to the ATP/ADP couple, as it were by a system of exchange rates.

How, then, do organisms make the ATP they require? The two major pathways that generate and regenerate ATP are called oxidative phosphorylation and photophosphorylation—driven by respiration and by light absorption, respectively. Both depend on the operation of chains

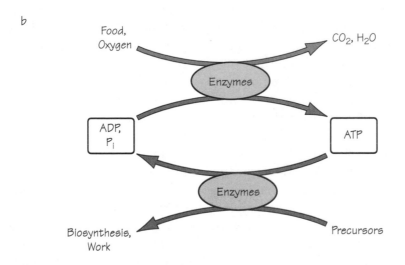

Fig. 4.4. ATP serves as the central energy currency in biological processes. (a) Adenosine triphosphate; the squiggles identify the "high-energy bonds" that underlie the function of ATP. (b) Getting and spending: How the ATP cycle couples energy-producing processes to energy-consuming ones.

of catalytic proteins embedded in a membrane: the inner membrane of mitochondria, the thylakoid membrane of chloroplasts or the plasma membrane of certain bacteria. Just how ATP is produced remained mysterious for many years, until it was discovered that the process is at bottom electrical. Both the respiratory chain, a bucket-brigade of proteins that mediate the oxidation of substrates by oxygen, and the analogous photosynthetic cascade, generate a current of protons across the membrane in which these proteins are inserted. These currents power ATP synthesis, and also serve directly as the source of energy for certain work functions. We can thus think of ATP and the proton current (more precisely, the proton potential) as alternative and interconvertible energy currencies. Some functions are paid for in the one currency, others in its mate. But this subject, like others in which membranes make an appearance, takes us across the border that separates molecular processes from cellular ones; we shall return to it in Chapter 5.

GENETIC INFORMATION

The unique mark of a living organism, shared with no other known entity, is its possession of a genetic program that specifies that organism's chemical makeup. The program has two essential and related features: first, it is "read" by the organism, and the instructions embodied therein expressed, second, it is replicated with high fidelity whenever the organism reproduces. Rare errors do occur during replication, these will be perpetuated henceforth and commonly alter the sense of the genetic program. Whether such mutations prosper or fail is determined by natural selection, exercised not upon genes but upon organisms. Mutations alter the genotype, selection evaluates the phenotype.

The chemical nature of the genetic material, deoxyribonucleic acid or DNA, was discovered just over half a century ago, in 1944, and its structure was published a decade later. The double helix quickly became the symbol of a new discipline, molecular biology, whose primary subject is the nature, replication and expression of genetic information (and, during the past decade, its manipulation for human purposes). The discovery of how genes are constructed and what they do ranks as one of the finest intellectual achievements of the twentieth century. Textbooks, obliged as they are to haul a great burden of factual detail, seldom convey more than the bare bones of this extraordinary episode. Curious readers will find much insight into the science and the personalities involved in articles by Stent and Chargaff, and particularly in H. F. Judson's lively and urbane account of the rise of molecular biology (8).

DNA Carries Genetic Specificity. The genetic functions of DNA grow directly out of its chemical structure (Fig. 4.5). DNA consist of two polynucleotide chains, wound about each other to make a helical duplex. The sequences of the two chains are complementary: where one has an adenine residue, the other has thymine; where one has guanine, its partner has cytosine. It is the weak but numerous hydrogen bonds between the complementary bases that hold the two chains together. This structure immediately suggests that genetic specificity, the "information" that distinguishes one gene from another, resides in the se-

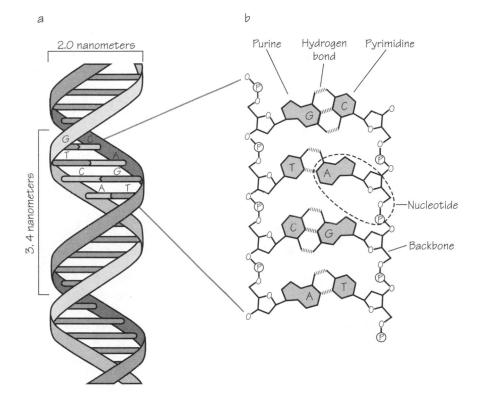

Fig. 4.5. DNA, the chemical basis of heredity. (a) The double helix. (b) The helix unrolled, to identify nucleotides, purine and pyrimidine bases, and the hydrogen bonds that hold the two helices together. A, adenine; G, guanine; C, cytosine; T, thymine.

quence of nucleotides. More precisely, since the sugar-phosphate backbone is the same in all DNA molecules, specificity must be conveyed by the sequence of purine and pyrimidine bases that project from that backbone (Fig. 4.5). Mutations consist of changes in the nature or the order of these bases. It was recognized from the first that the duplex structure of DNA implies a simple mechanism by which the base sequence can be replicated exactly whenever the cell divides: let the two chains unwind and separate, and let each sequence of bases serve as the template for the construction of a new chain complementary to itself. We now know that this is, in principle, how DNA is replicated, a process mediated by the enzyme DNA polymerase. The actual mechanism, however, is rather more complex for two reasons. First, the two complementary DNA chains run in opposite directions ("antiparallel"), and must therefore be copied separately by different mechanisms, and second, because special devices are required to ensure that the copying

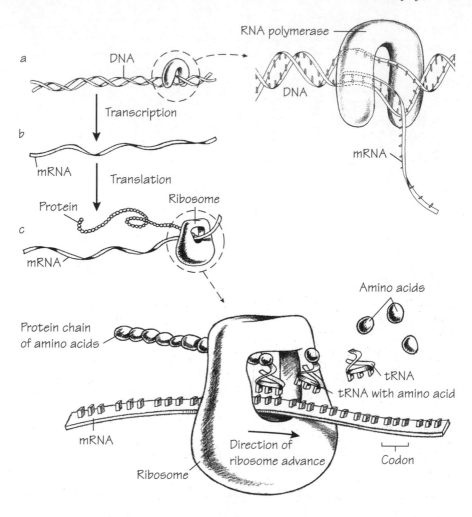

Fig. 4.6. The secret of life: a molecular perspective. (Upper left) How genetic information is replicated, transcribed into mRNA and translated into proteins. (Upper right) Transcription. RNA polymerace is the enzyme that links together nucleotides in the order that complements the DNA strand being transcribed. (Bottom) Translation and function of ribosomes. Ribosomes link amino acids into a growing chain; transfer RNAs present the activated amino acids; and messenger RNA specifies the order, or sequence, of the amino acids.

process be as nearly as possible error-free. It bears repeating here that, except for a handful of viruses, DNA is the universal genetic material, and its replication constitutes the mechanism by which instructions have passed from one generation to the next for well over three billion years.

DNA makes RNA makes Protein. Even before the nature of the genetic material was understood, it had been recognized that genes somehow determine the properties of proteins, and that, as a rule, each gene supervises one particular protein. This is how genes determine traits. But the double helix did not in itself explain just how a gene determines the structure and function of the corresponding protein. That had to be discovered step by step, and most of the steps had to be imagined before they could be observed. Messenger RNA, transfer RNA, ribosomes, even that universal set of twenty amino acids—each of these represents a leap of the imagination, disciplined by reflection and by many tedious hours at the laboratory bench.

By now, the upshot is taught to every high school student (Fig. 4.6). Genes determine the structure of proteins because the sequence of bases in DNA designates the sequence of amino acids in the corresponding protein. We speak of this as the transfer of "information," a specific order of symbols, from DNA to protein. Generally speaking, the identity and function of a protein is specified entirely by its amino acid sequence; the linear amino acid chain folds up spontaneously into a particular three-dimensional shape, upon which that protein's function then depends. In bacteria, the sequence of amino acids can be mapped point by point upon the sequence of nucleotides in the corresponding gene. Eukaryotic genes are more complex, containing stretches of DNA that do not encode amino acid sequences. These inserts are excised when the information carried by the gene is expressed; their function is still debatable. The set of rules that specify which sequence of bases specifies each of the standard suite of twenty amino acids is known as the genetic code. It is a nonoverlapping triplet code: three consecutive bases specify each amino acid (AAA stands for phenylalanine; AAT for leucine), and it is one of the universal commonalities of living organisms. The table of codons that specifies which triplets stand for each of the twenty amino acids has no known theoretical basis, and had to be worked out by experiment.

Any alteration in the sequence of DNA, once replicated, is inherited henceforth; that is the chemical basis of mutation, and therefore of much of the genetic variation within populations. Such a sequence change commonly (but not always) elicits a corresponding change in the amino acid sequence specified by that gene, and many of these changes (but by no means all) entail consequent alterations in the shape and function of the resulting protein. Mutations take many forms: they may result from rare errors of replication or from such environmental insults as radiation; they may be restricted to a single base or entail the loss of a chromosome segment bearing many genes. But they have this in common: in nature, mutations occur at random. Variation is generated by unpredictable, chance events.

Even though the mechanisms by which the nucleotide sequence of DNA specifies the amino acid sequence of protein are complicated, the principle is quite simple (Fig. 4.6). Briefly, the sequence of one DNA strand is first transcribed into a more portable form, a single-stranded RNA molecule that moves off into the cytoplasm. The production of this messenger RNA (mRNA) is the task of an RNA polymerase, and the sequence of bases in mRNA is complementary to that of the DNA, base by base (except for excised introns). Messenger RNA constitutes a tape that is then read, and translated into protein language, by ribosomes. Complex structures in themselves, ribosomes stabilize the line-up of amino acids, one for each nucleotide triplet as specified by the table of codons, and then zip them together. Ribosomes, together with an energy-donating molecule related to ATP, do chemical work but they contribute no information: the sequence in which the amino acids are lined up is wholly specified by the mRNA tape. But the molecular mechanics are dauntingly complicated. For instance, mRNA cannot recognize amino acids directly. It recognizes amino acids attached to an adaptor, a small RNA molecule called transfer RNA (tRNA), which lines up on the mRNA codon with one end and with the other presents the correct amino acid to the ribosome (Fig. 4.6).

The result is encapsulated in the familiar phrase, "DNA makes RNA makes protein." For "make" read "specifies the sequence of," for in fact the manufacture of all these molecules is the responsibility of complex enzymes such as RNA polymerase or the ribosome. The same is true for DNA: one generation of DNA molecules specifies the sequence of the next generation, but it is quite inaccurate to speak of DNA as a self-replicating molecule. An enzyme called DNA polymerase is required to replicate any DNA sequence. We can summarize the flow of information by the linear sequence

$$DNA \quad \rightarrow \quad RNA \quad \rightarrow \quad Protein$$

But the production line feeds back upon itself: at every stage, certain proteins specified or coded by the DNA are required to carry out the chemistry.

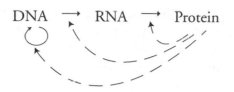

DNA → RNA → Protein

Any pathway of communication is prone to error. It has been cal-culated that the basic process of protein synthesis sketched above would insert the wrong amino acid at least once in every 200 instances. That is not nearly good enough, for many proteins are present in but a few copies per cell and an error can be fatal. The actual frequency of error is closer to one in 4,000, attained with the aid of ancillary proteins that "proofread" the mRNA transcript and monitor the performance of ri-bosomes. Proofreading is even more critical during DNA replication, and this ensures the low normal mutation rate of one base pair in a million for each replication cycle, or even less.

The Central Dogma. The transmission of genetic instructions from DNA to proteins has another crucial feature: it is unidirectional. Information flows from DNA to RNA (and, under special circumstances, in the reverse direction also), and thence to protein, but never from protein to nucleic acids. Let us make this statement more concrete. A change in the sequence of a gene, a mutation, is reflected in a change in the amino acid sequence of the protein coded by that gene. It is also in-herited by copies of that gene. But a change in the amino acid sequence of a protein (elicited, perhaps, by some chemical procedure) will not be inherited or preserved in any DNA sequence. Even if the altered protein were an enzyme that replicates DNA, its catalytic rate or accuracy may be affected but there cannot be a specific (reproducible) effect upon the sequence of the DNA replicas.

The principle, that information flows unidirectionally from nucleic acids to proteins but never in the reverse direction, was formulated by Crick (9) and dubbed the "central dogma of biology." Like the laws of thermodynamics, it rests not on direct evidence but on the absence of any known exception. The central dogma represents, at the molecular level, the well known biological principle that acquired characteristics are not inherited (which, incidentally, does have exceptions; see Chapter 7). Sequence information is only accessible by way of the genes. Any change must be effected at the DNA level, normally through mutation, and the biological meaning, or value, of this alteration is judged at the protein level, the level of function as expressed in the organism as a whole. There appears to be no way in which changes in the environment can call forth specific gene mutations. No matter how pressing the need, change must wait upon chance mutations.

PROTEIN FOLDING: GIVING MEANING TO MESSAGE

Genetic information flows in linear fashion from the sequence of bases in DNA to that of amino acids in proteins. The parallel with letters and words is inescapable, reinforced by the vocabulary of metaphorical terms employed in the trade (messages are encoded on a tape, proofread, transcribed and translated, sometimes spliced, and they make sense, antisense or at worst nonsense), and the quantity of information transmitted can be estimated with the aid of algorithms derived from wartime researches on the fidelity of communications. But sequences are just strings of symbols without intrinsic significance. At the end of the day, the object of the genetic exercise is to specify the shape of a protein that performs a biological function. Folding of the nascent polypeptide, the transition from one dimension to three, is what gives the message its meaning.

The principle that governs this transition was discovered by Christian Anfinsen and his colleagues forty years ago. Working with the small enzyme ribonuclease, they found that the "denatured" (i.e., unfolded) protein would spontaneously refold itself into the native conformation in less than a minute, concurrently recovering catalytic activity. Aside from a suitable buffer, no energy source or any other substance was required. Experiments of this kind have been performed with many other proteins as well, all leading to the same general conclusion: the amino acid sequence is fully sufficient to determine that protein's three-dimensional configuration, and also its biological activity. It follows that shape and function are already implicit in the coding sequence of the corresponding gene. In principle, then, the forces responsible for shaping the protein are a matter of physics and chemistry alone, and the shape of the protein should be predictable from its sequence. This cannot yet be done; the protein folding problem may be the largest gap remaining in the fabric of classical biochemistry.

Before proceeding, we must note a complication. In the living cell, folding of nascent polypeptide chains is commonly assisted by a class of proteins dubbed chaperonins (after their function of preventing inappropriate interactions between complementary surfaces). In many instances, large polypeptides in particular, the intervention of chaperonins is essential for quick and accurate folding (or refolding). But this does not contradict the generalization that all the information required to shape a protein is contained in its amino acid sequence. Chaperonins are catalysts, and no more than that. Mutations in the protein's coding sequence affect its form and function; mutations in the chaperonin do not.

Though we cannot presently predict the shape produced by the folding of some given amino acid sequence, the principles of protein architecture supply signposts towards that goal. First, every protein is a unique construct, made up of such recurring elements as α-helices and β-sheets; these arise predictably and quickly from short polypeptide stretches containing one or two dozen amino acids. Second, the number of possible configurations is greatly restricted by the requirement that hydrophilic amino acids be exposed; on the outside whereas hydrophobic ones must be buried in the interior, moreover, the correct configuration is one that maximizes hydrogen bonds between adjacent amino acid residues. Proteins are notoriously labile, and a single misplaced residue may be sufficient to destabilize the entire structure. Finally, one can easily calculate that protein folding cannot take place by successive trial of possible conformations; there is nowhere near enough time for random searching. Instead, proteins fold by tracing a more or less unique pathway, involving both global and local interactions. The identification of these kinetic routes, which may differ from one protein to another, is presently the focus of intensive research. A good amino acid sequence is evidently one that folds into the functional form, and does it quickly and correctly every time.

Why has protein folding proven such an intractable puzzle? Perhaps it is because it poses, in microcosm, the larger question of biological morphogenesis. Many parts, each known in detail, come together to make a far larger entity whose form is but distantly related to that of its elements. Like a bicycle or a cell, every protein is a whole greater than the sum of its parts. The form of a protein is an emergent property, not easily predicted from a knowledge of its chemical makeup.

Keeping Control: Regulation and Homeostasis

Bacterial cells (e.g., *E. coli*), growing on a medium that consists only of inorganic salts and glucose, must synthesize all their constituents from scratch. The many chemical reactions that this entails are evidently closely regulated, for normally the cells avoid both shortfalls and excess production of constituents. Now, let us add to the medium a single essential metabolite, the amino acid tryptophan. The cells respond with admirable frugality and promptness to shut down the production of endogenous tryptophan: enzymes in the pathway of tryptophan biosynthesis are inhibited, and production of the enzymes themselves is halted. Both responses are executed by the binding of tryptophan to a cytoplasmic protein which, in effect, measures the availability of the metabolite.

In cellular metabolism, homeostasis is the rule: the internal environment remains constant within narrow limits. The concentrations of ions, the pH, and the osmotic pressure remain essentially invariant. The rates of most metabolic reactions, and such global processes as cell division, are subject to control, both inhibitory and stimulatory. This is accomplished by an intricate network of signals, parallel to and interwoven with the metabolic web, whose object is to convey information about the status of some parameter, and this information is used to adjust the rate of at least one chemical reaction. The sense of "information" in the present context is not the same as that of the genetic information discussed in an earlier section, but the various kinds of information have a key feature in common: if energy is the power to do, information is the power to direct what is done (10).

Some of the molecules whose concentration conveys a message or signal are themselves metabolic intermediates, as tryptophan is. Others serve no function other than to carry information (e.g., calcium ions or cyclic AMP). Many are small, others are proteins or portions of proteins (such as the leading sequences that direct certain proteins to a particular location). Some signals, addressed to the cell as a whole, speak to surface receptors, others coordinate intracellular operations. Some regulate the activity of enzymes, others control the expression of genes. The overall purpose of this network of signals is plain enough: stability of all operations, the efficient use of resources and appropriate responses to changes in the environment. But the mechanisms employed are subtle, surprisingly complex, and exceedingly diverse. The diversity of regulatory devices is far greater than that of the processes which are subject to regulation.

One recurring principle of regulatory mechanisms is that association of a protein with another molecule (or "ligand") causes a change in the protein's shape, and the altered shape is accompanied by a change in its activity. The ligand may be an ion, a small signaling molecule or another protein, and the change in activity may refer to catalytic rate or to the capacity to associate with another macromolecule. The regulation of tryptophan biosynthesis, mentioned above, is a case in point. One of the enzymes early in the pathway binds tryptophan at a special regulatory site; this changes the protein's shape, and its enzymic activity is diminished in consequence. Tryptophan thus exerts feedback control over its own production. In other cases, the regulatory change in conformation is brought about by the chemical modification of a protein: many enzymes are turned on (or shut off) by the transfer of a phosphoryl group from ATP in response to some environmental stimulus. There are cascades of such "kinases," one switching on the next, and each potentially the target for a signal. The common conception of proteins serving either as catalysts or as structural elements is incomplete: for many purposes it is more useful to emphasize their roles as computational elements in information-processing networks (11).

The regulation of gene expression is of particular and intense interest at the present time, whether, and at what rate, a given gene is transcribed and translated is almost always subject to control by an amazing variety of regulatory devices. The most familiar of these, and the first to be elucidated (by François Jacob and Jacques Monod, who won a Nobel prize for their work in 1965), takes the form of regulatory proteins that bind to a region of DNA upstream from the coding sequence (the "gene" proper), and thereby either repress transcription or promote it. The target is commonly RNA polymerase: the regulatory protein may block access of the polymerase to the point where transcription begins, or alternatively may promote that association. In the case of tryptophan biosynthesis, the production of the enzymes of tryptophan synthesis in *E. coli* is inhibited when tryptophan is present (Fig. 4.7). Tryptophan binds to a cytoplasmic repressor protein; this promotes the repressor's association with a particular DNA sequence, thereby diminishing transcription of the entire set of tryptophan genes, which are clustered together in tandem. But that is not the whole story. A second level of control, called attenuation, depends on the translation of the message: in the presence of free tryptophan ribosome movement is retarded, which tends to feed back upon transcription so as to abort it. The use of multiple, unrelated, and redundant regulatory devices is quite typical; the object of the exercise is not simplicity and elegance, but to make both the activity and the production of the enzymes of tryptophan biosynthesis exquisitely sensitive to the supply of cytoplasmic tryptophan.

Control circuits, particularly those seen in eukaryotic cells, are more elaborate than the processes which are regulated, and commonly involve many more components. The regulation of gene expression in eukaryotic cells depends on the same principles as those outlined above, i.e., binding of regulatory proteins to regulatory DNA sequences, and the modulation of this binding by various signals. But the "switches" are more sophisticated, so as to integrate input from half a dozen directions

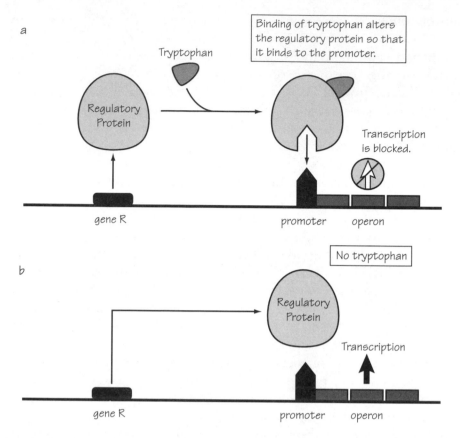

Fig. 4.7. Regulation of enzyme synthesis: the case of tryptophan. (a) Gene R produces a regulatory protein that binds to tryptophan; this alters the conformation of the protein so that it binds to the promoter, blocking transcription of the structural genes downstream. (b) In the absence of tryptophan the regulatory protein fails to bind and transcription proceeds. (Attenuation not shown.)

at once. In consequence, a large share of the cell's genetic information must be devoted to the production of regulatory elements. The gain is an elaborate network of controls that links transcription to other cellular functions, weaving them all into an integrated web.

SELF-ASSEMBLY: MOLECULES INTO STRUCTURES

Some molecules carry out their tasks singly; many enzymes are loners, as are transfer RNAs. But the majority of biological functions depend upon larger and more elaborate structures, composed of several individual polypeptide chains (as is hemoglobin) or assembled from a number of diverse macromolecules. DNA famously consists of two intertwined antiparallel strands with complementary sequences. Ribosomes are made up of some 50 distinct proteins and three (sometimes four) sorts of RNA, distributed between two large subunits; the F_1F_0-ATPase that pumps protons across prokaryotic membranes (Chapter 5) is not much smaller. Such multimolecular complexes commonly arise by the spontaneous association of their constituent molecules, with little or no input of energy or additional instructions. To a first approximation (and bear in mind how many ambiguities can shelter under this bland cover), all the information required to produce the shape and function of the complex is supplied by its molecular constituents, much as the pieces of a jigsaw puzzle determine the complete pattern. Furthermore, since the shape of each protein is a function solely of its sequence, which is in turn encoded by a gene, the genome implicitly encodes the form and function of the assembled complex.

The domain of self-assembly is large, extending over many of the familiar cellular entities and organelles. The phenomenon was uncovered in the early fifties in the course of research on the nature of viruses. Tobacco mosaic virus can be dissociated into RNA (in this instance, the genetic material) and the protein subunits of the outer shell; the purified macromolecules reassociate readily to reconstitute infectious virus particles. Bacterial flagellin assembles into flagellar filaments, and the surface proteins of certain Archaea into the characteristic S layers. Eukaryotic cells feature the spontaneous association of tubulins into microtubules, of actin into microfilaments, and of myosin into "thick filaments." Nucleosomes assemble from histones and wrap themselves in DNA; the nuclear membrane pores, which mediate passage into and out of the nucleus, are likewise products of self assembly. Nor is the province restricted to protein assembly. Phospholipid bilayers, the basic structure of biological membranes, form spontaneously when phospholipids are suspended in water. The most stable configuration is a closed vesicle, with all the polar headgroups in contact with water and the nonpolar tails huddled together in the interior of the lipid membrane (Fig. 4.2).

We can cast the net still wider if we emphasize, not the absence of input from external sources of energy or information, but the degree to which molecular structures inform higher levels of cellular organization. Protein folding, fully specified by the primary sequence of amino acids, is a case in point. Portions of that sequence can even serve as an address label, directing the protein to its destination in the cell: proteins destined for secretion, for insertion into membranes or for incorporation into mitochondria, nuclei, and other organelles are identified by features of the sequence or of their shape. Recent reports indicate that the mitotic apparatus can be considered a self assembling entity, and so can the eukaryotic nucleus as a whole.

The idea that biological organization is fully determined by molecular structures is popular, seductive, potent and true up to a point—yet fundamentally wrong. Many scientists cling hopefully to Lederberg's dictum of thirty years ago: "The point of faith is this: make the macromolecules at the right time and in the right amount, and the organization will take care of itself" (13). But this faith is too simple to suit modern knowledge. It disregards the fact that the cell as a whole is required to create the proper environment for self-assembly to proceed. Furthermore, both prokaryotic and eukaryotic cells make sure to control self-assembly, so that it takes place only as part of a larger purpose. Here we stand once again on the border that divides biochemistry from cell biology, and the present chapter from subsequent ones.

THE UNITY OF BIOCHEMISTRY

A chemist investigating the composition of living organisms comes away with two conclusions. First, even the simplest cell is an exceedingly complex mixture containing thousands of different molecules, whose proportions remain essentially invariant generation upon generation. Second, from the chemical viewpoint, rabbits and grass are very much alike, and their molecular constituents comprise but a tiny fraction of the structures known to chemistry. All cells contain virtually the same set of small molecules—amino acids, sugars, sugar phosphates, nucleotides, dicarboxylic acids, perhaps a hundred in all, dissolved in water, which makes up as much as nine tenths of the total mass. Potassium ions always serve as the chief intracellular electrolyte, even though sodium ions predominate in the environment. And their macromolecules are all made to the same plan. The myriad of known proteins are all constructed from a common set of twenty amino acids, linked into unbranched chains by a coupling called the peptide bond. A common set of five nucleotides accounts for all the many molecules of RNA and DNA, except for a sixth nucleotide found only in a handful of viruses. Diversity among the lipids and carbohydrates is somewhat greater, but the repertoire is still quite limited.

This should not be construed to mean that all living organisms are chemically identical. The unity of biochemistry has limits: steroids occur in eukaryotes but seldom in prokaryotes, cell wall peptidoglycan is confined to the Eubacteria, Archaea make their membranes of ether-linked lipids rather than ester-linked ones, pigments and pheromones mark particular species. But what chiefly determines the chemical identity of each species is the set of amino acid and nucleotide sequences that define its particular complement of macromolecules. There are 20^{100} ways of stringing together twenty amino acids into a protein 100 amino acids in length, leaving ample room for diversity. When one compares sequences, the macromolecules of even closely related species, though plainly similar, are almost always distinct, and some degree of variation can often be found even between individuals. We are looking at a unity of design, with a wide range of variations in the details. Unity extends beyond the chemistry of molecules to the construction of large complexes comprised of dozens of molecular elements, and to the articulation of metabolic pathways. In both areas one can readily discern a core set of processes common to all living organisms with comparatively minor variations, and a wide periphery of increasing diversity.

Protein synthesis exemplifies the core, all the way from the manner in which nucleotide sequences specify amino acid sequences to the machinery that turns out protein chains. We have here a universal constellation of organelles and molecules, collaborating in a fixed order to a common end: RNA polymerases, messenger RNAs, transfer RNAs, enzymes that activate amino acids and attach them to RNA, ribosomes, and a selection of co-factors. To be sure, the molecular components extracted from diverse organisms are similar rather than identical. Ribosomes of bacteria and eukaryotes differ substantially in size, composition and sensitivity to inhibitors, and the diverging sequences of both RNAs and proteins have been used to trace the phylogeny of the organisms from which they come (Chapter 3). Yet the general nature of the system, what it does and how it is articulated, is essentially universal; like a sedan and a pickup, they display significant variations upon a standard design.

The most compelling instance of biochemical unity is, of course, the genetic code. Not only is DNA the all but universal carrier of genetic information (with RNA viruses the sole exception), the table of correspondences that relates a particular triplet of nucleotides to a particular amino acid is universal. In all organisms, GUU stands for valine, UAU for tyrosine. There are exceptions (in yeast mitochondria, for example, the standard stop codon UGA is read as tryptophan), but they are rare and do not challenge the rule.

Turning now to the metabolic web, that network of linked reactions by which molecules are produced, degraded and interconverted, unity and diversity are more evenly balanced. Take glycolysis, the route by which glucose is broken down anaerobically with the concurrent production of useful energy. The pathway is very widespread among

Eubacteria and Eukarya, and was at one time considered to be one of life's universals. That is clearly not the case: it is absent from photosynthetic bacteria, and it appears, from the Archaea. Likewise, the pathways of amino acid biosynthesis are widespread but not universal: bacteria and fungi employ different pathways for the generation of lysine. As to energy generation, a wide range from unity to diversity is on display. Adenosine triphosphate, pyridine nucleotides, flavins and proton currents are universal features. Electron transport proteins with heme prosthetic groups are found almost everywhere; these clearly make up a family of related proteins serving a common function. But the enzymes that mediate access to environmental sources of energy differ considerably among organisms that rely on light, preformed organic matter or inorganic processes.

The unity of biochemistry was first clearly enunciated by A. J. Kluyver and the Dutch school of microbial physiologists half a century ago, and surely represents one of the most profound insights into the nature of life. We take it as incontrovertible evidence that all life on earth is of one kind, descended from a common ancestor. The universal core represents our common inheritance, the diversified periphery displays subsequent divergence by variation and natural selection. *E pluribus unum* (from diversity, unity) could stand as a motto for biology; provided we hasten to add "and from unity, diversity."

MOLECULAR EVOLUTION: THE MARK OF THE TINKERER

The molecules of life make up a minute fraction of the organic substances known to chemists: why just these and not others? One possible explanation is chemical necessity: these molecules, and they alone, have the properties required to perform their function. No one doubts that the structures and properties of biological molecules do suit the roles they play, but there is no reason to believe in a unique fit, and much evidence that contradicts the notion. What adenosine triphosphate does, guanosine triphosphate could do as well; the fact that ATP serves as the universal energy currency while GTP performs specialized tasks (in protein synthesis and cell signaling) is not explained by the difference in chemical structure. By the same token, and despite diligent effort, no one has discovered a persuasive chemical connection between any particular triplet of nucleotides and the amino acid that this triplet codes for. A far more plausible hypothesis, universally accepted by contemporary biologists, holds that the twenty amino acids and the five nucleotides, ATP and NADH and the rest of the core metabolites, were part of the endowment of that ancestral cell line from which all contemporary organisms descend. These molecules are so deeply embedded in the fabric of cellular biochemistry that any alteration would be lethal. Molecules that are not part of the common core are peripheral in the sense that they arose subsequently in one lineage or another.

Molecular Historiography. Of the origin of the central metabolites only the faintest vestiges survive; but macromolecules preserve in their very structure a record of their evolution. In the mid-sixties, when the amino acid sequences of proteins first became available, it was realized that the sequences of whale hemoglobin and porpoise hemoglobin were almost identical, while that of horse hemoglobin was considerably different. In general, the more distant the relationship between the parent organisms, the more numerous are the differences in sequence, suggesting that all vertebrate hemoglobins are descended by progressive modification from an ancestral form of hemoglobin. Unexpectedly, it turned out that differences accumulate at an approximately constant rate, such that the extent of the difference can serve as a measure of the time elapsed since any two proteins diverged from their common ancestor. One can use macromolecules as markers to track the evolution of organisms, supplementing the notoriously patchy evidence of the fossil record; alternatively, one can trace descent with modification of related proteins, such as those that mediate the uptake of metabolites.

The new molecular technology, particularly the development of methods for the rapid sequencing of minute quantities of proteins or nucleic acids, have transformed the stodgy sciences of taxonomy and phylogeny. We can now examine evolution at the molecular level, where genetic variations arise, and in consequence the history of life is accessible as never before. That the ancestry of humans diverged from that of the great apes a mere 5 to 6 million years ago is now a staple of the textbooks, but was derided as an outrageous heresy when the late Allan Wilson first inferred it from amino acid sequences three decades ago. Now the controversy swirls over DNA sequences: the small mitochondrial DNAs diverge so quickly that they can be used to track recent evolutionary events, such as the emergence and dispersal of modern humans, which appears to have begun as little as 200,000 years ago. At the opposite extreme, ribosomal RNAs are strictly conserved, they evolve so slowly as to preserve relationships that go back billions of years. Contemporary bacterial phylogeny is based very largely on ribosomal RNA sequences (Chapters 3 and 8). This approach also underpins the profound discovery that all living organisms can be subsumed under three domains: Eubacteria, Archaea and Eukarya (Fig. 3.2). The same methods now make it possible to classify bacteria that cannot be cultured and have never been seen by any microscopist, among them some that branch off at the very base of the tree of life. Now whole genomes tumble like ninepins before the assault of the robotic sequencers. I confess to some skepticism concerning the promised benefits for human health and happiness, but no one can doubt the impact of the new information on our understanding of biological evolution.

One immediate benefit is the resolution of an old conundrum. As Monod pointed out in his provocative book, *Chance and Necessity*, the structure of every protein is random "in the precise sense that, were we to know the exact order of 199 residues in a protein containing 200, it would be impossible to formulate any rule, theoretical or empirical, enabling us to predict the nature of the one residue not yet identified by analysis" (14). The same, no doubt, holds for nucleotide sequences in RNA and DNA; yet all organisms take great pains to ensure the accurate production and reproduction of these apparently meaningless strings of symbols. The resolution of the paradox is found in the historical nature of all the informational macromolecules. One can predict the identity of the next amino acid quite successfully from a knowledge of the sequence of a related protein in either the same or another organism. It is not chemical law but biological function that governs the sequence, and its progressive transfiguration can often be traced by molecular historiography. Incidentally, the fact that the function of a protein depends upon its form, rather than directly upon its amino acid sequence, helps one understand the many instances in which three-dimensional structure is better conserved than the sequence. For example, the sequence of the bacterial division protein FtsZ is but distantly related to that of tubulin, but when one compares global shapes the homology leaps to the eye (Chapter 6).

When one looks at macromolecules as historical documents, the traditional concepts of homology and analogy take on new meanings. Homologous organs are those that share a common ancestry; for example, the bat's wing is homologous to the human hand and forearm even though these organs serve quite different functions. By contrast, the wings of bats and butterflies are quite unlike in structure and genesis, and would be termed analogous. What is new is that the molecules responsible for basic cellular processes are often highly conserved, revealing deep homologies that were obscured by subsequent modification at the organismic level. Thus, the proteins that underpin cell division in yeast and animal cells are not only recognizably homologous but may be functionally interchangeable; the genes that encode the proteins that effect patterning of the body plan in insects and vertebrates have basic features in common, even though the pathways of embryogenesis are grossly different. The meaning of these discoveries is just being spelled out (Chapter 9). They suggest, at the very least, that the continuous generation of evolutionary novelty rests on a solid foundation of conservatism and constraint.

Sources of Hereditary Variation. The molecular basis of simple mutations, creeping at a petty pace from nucleotide to nucleotide, was worked out early in the development of molecular genetics. Since then, the repertoire of hereditary change has expanded prodigiously, making our conception of the genome far more fluid. Single-base mutations, recombination, and deletions remain in the picture. But the major source of evolutionary novelty now appears to be gene duplication, followed by progressive divergence; the many instances of homology among proteins of widely different functions can be understood on this basis. Genomes accommodate, and often preserve, various mobile genetic elements (viruses, transposons, plasmids), and these make possible the occasional transfer of a gene across a large taxonomic distance; among them are bacterial genes for glycolytic enzymes that seem to have come from eukaryotes, and the superoxide dismutase of *Entamoeba histolytica*, which appears to be of bacterial provenance. And at a much higher level of order, endosymbiosis provides a major source of genetic variation: mitochondria and chloroplasts quite certainly represent relics of such episodes. In the former case, most of the genetic instructions of the symbiont were subsequently transferred to the nuclear genome of the eukaryotic host, but they still preserve traces of their prokaryotic origins.

Between Selection and Tolerance. Protein sequences diverge at a relatively constant rate over hundreds of millions of years; among the globins, one amino acid is replaced by another every four million years on the average. This is a great boon to students of phylogeny, providing them with a rough "molecular clock." But the constancy of the rhythm raises serious questions concerning the role of natural selection in guiding the "design" of proteins. If every sequence change were adaptive and delivered improved performance, one would hardly expect replacement to occur at a constant rate. The clock-like behavior suggests, instead, that most amino acid replacements are neutral in their consequences, and that natural selection plays but a minor role in protein evolution. The debate between neutralists and selectionists has not been altogether settled, but it has drifted towards the middle ground (15). It is clear that amino acid residues (or short sequences) that are directly involved in that protein's biological function tend to be highly conserved; an example would be the amino acids whose conjunction creates an enzyme's catalytic site. The great majority of other residues, however, seem to matter much less and are free to vary within limits imposed by the stability or the folding of the protein as a whole. Evolution, as François Jacob pointed out long ago is not an engineer or a designer but a tinkerer.

A marked degree of error tolerance seems to characterize some genomes as well. Bacterial genomes, which consist almost wholly of useful sequences that code either for amino acid sequences or for regulatory elements, give the impression of being highly streamlined. Eukaryotic genomes, especially those of higher organisms, are quite otherwise. They tend to be much larger than those of prokaryotes, and carry a substantial load of apparently non-functional DNA. In fruit flies, the fraction of DNA that codes for something identifiable is about a third; it is a tenth or less in humans, and less than 1 percent in certain flowering plants. The remainder is commonly dismissed as "junk DNA," but whether the epithet is justified remains in some doubt. It is also quite unclear why eukaryotic genes are commonly interrupted by "introns," noncoding stretches of DNA, which are transcribed but subsequently spliced out of the messenger RNA. Informed opinion tentatively holds that introns are not remnants of the early evolution of genomes, but the "fossil" remains of much more recent insertion of mobile genetic elements such as viruses; here again, open-minded skepticism may be the beginning of wisdom. Whatever the nature and origin of supernumerary DNA, genomes appear to be even more variable, or less constrained, than the products they specify.

It would, presumably, be possible in principle to shed the burden of non-functional DNA, which must impose some cost on the organism, but the cost of streamlining may be the greater. Besides, genetic sloth may bring its own reward, allowing a degree of evolutionary flexibility that is denied to prokaryotes. Could there be a connection between the large, error tolerant genomes of eukaryotes and their self-evident capacity for morphological and physiological diversification? However, much my molecular colleagues may wish it were otherwise, biology is an irremediably historical science; even its molecules are infected with contingency. One is tempted to say, with apologies to Theodosius Dobzhansky, that little in biochemistry makes sense except in the context of history.

5

A (ALMOST) COMPREHENSIBLE CELL

"I am middle aged now, but in the autumn I always seek for it again hopefully. On some day when the leaves are red, or fallen, and just after the birds are gone, I put on my hat and an old jacket, and over the protests of my wife that I will catch cold, I start my search. I go carefully down the apartment steps and climb, instead of jump, over the wall. A bit further I reach an unkempt field full of brown stalks and emptied seed pods."

Loren Eiseley (1)

WHAT'S SPELLED IN THE GENES
THE DYNAMIC ORDER OF METABOLISM
THE LIMITS OF SELF-ASSEMBLY
LEGACY OF A REVOLUTION
READING THE MICROBIAL MIND
HOW STRUCTURED IS THE CYTOPLASM?
AND THEREFORE, WHAT?

There is more than one way to seek the secret of life, and many places where it may be found—or, at least, looked for. Loren Eiseley, naturalist and poet, musing in a suburban field among "seeds and beetle shells and abandoned grasshopper legs" found "something that is not accounted for very clearly in the dissections to the ultimate virus or crystal or protein particle;" and surmised that "life is not what it is purported to be" (1). About the same time, half a century ago, microbiologists became persuaded that if we would discover the nature of life, we should concentrate our efforts on the bacterial world, *Escherichia coli* in particular. Our quest and Eiseley's are the same, but in the simplest unicellular organisms the secret presents itself in its most elemental form, and the outlook is correspondingly more hopeful.

The basic procedure has been performed thousands of times in laboratories the world 'round. A few cells of *E. coli* (in principle, a single one will do) are removed from an earlier culture with the aid of a sterile metal loop, and transferred to fresh sterile growth medium. The medium is a solution of inorganic salts including phosphate, sulfate, ammonium, and potassium ions; a number of trace metals; and a pinch of glucose. The flask is then placed in an incubator, preferably on a shaker. Next morning the glucose has been consumed, and the medium swarms with cells, billions per milliliter, all identical with those of the inoculum. Here in microcosm is all the mystery of life. As the cells grow and multiply, they marshal vast numbers of atoms into a new and highly ordered formation, converting energy into organization. Each cell is a swirling vortex in the stream of matter and energy yet maintains its identity, a particular pattern in space and time. Its activities and behavior are purposeful, inasmuch as they contribute to the cell's survival, growth, and multiplication. As they multiply, the cells reproduce their own kind and no other; like begets like. And cells of *E. coli* never arise spontaneously; each one comes from a previous one. When (or perhaps if ever) we truly understand how and why even the simplest cells perform these marvelous feats, we will have come a very long way in the quest for the secret of life.

Project *E. coli* grew out of the researches of André Lwoff, François Jacob and Jacques Monod, initiated in Paris just before the Second World War and continued intermittently under the German occupation. It took on a burgeoning life of its own, not because *E. coli* is of special interest (it is a minor, and usually harmless, denizen of the human intestine), but because it proved to be a particularly convenient experimental organism. And the more that was learned about it, the more new questions could be asked, and the more investigators were drawn into the field. The program necessarily began with studies on the growth, nutrition, metabolism, genetics, and behavior of living cells. However, it soon took a markedly analytical turn, in keeping with a narrower conception of the objective: to account for all the cell's activities in terms of the structures and interactions of its molecular constituents. That is not quite the same thing as seeking to understand the living cell, but it makes a more attainable goal. Research at the level of cells and organisms has not been abandoned, but is now decidedly outside the main stream; most current research centers on limited and specialized questions, such as the regulation of an individual gene or the molecular mechanism of a particular reaction. Taken on its own terms, the effort has been resoundingly successful. You need look no farther than the massive compendium entitled *Escherichia coli* and *Salmonella* (2): with 2,800 double-column pages and more than 20,000 references it records almost everything known about these two closely related organisms, and represents something like an encyclopedia of *E. coli* life processes. The full sequence of the *E. coli* genome has just been determined, a few years hence, when every gene product has been assigned a function, our cup will indeed be filled to overflowing.

We microbiologists have every reason to be proud of our collective achievement, but the editors themselves hint at feeling (as I do) surfeited and unsatisfied all at the same time. Like Eiseley, one senses that something is not accounted for very clearly in the single-minded dissection to the molecular level. Even as the tide of information surges relentlessly beyond anyone's comprehension, the organism as a whole has been shattered into bits and bytes. Between the thriving catalog of molecules and genes, and the growing cells under my microscope, there yawns a gulf that will not be automatically bridged when the missing facts have all been supplied. No, whole-genome sequencing won't do it, for the living cells quite fail to declare themselves from those genomes that are already in the databases. We presently know something of half the genes of *E. coli* and the products they encode, but none of this information hints at a cylindrical cell with hemispherical caps. The time has come to put the cell together again, form and function and history and all.

Biochemists insist, rightly, that when one takes cells apart one finds nothing but molecules: no forces unique to life, no cosmic plan, only molecules whose writhings and couplings underlie and explain all that the cell does. Thus Max Perutz, reflecting on the mechanisms that allow *E. coli* to detect and swim towards a source of nutrients, found nothing that could not be "reduced to chemistry" (3). I share the commitment to a material conception of life, but that makes it doubly necessary to remember that before the cells were taken apart—as long, indeed, as they were alive—they displayed capacities that go beyond chemistry. Homeostasis, purposeful behavior, reproduction, morphogenesis, and descent with modification are not part of the vocabulary of chemistry but point to higher levels of order. Even as the catalog of small parts approaches completion, the transition from molecular chemistry to the supramolecular order of the cell emerges as a prodigious challenge to the imagination. Make no mistake about it: here we touch, if not the very secret of life, at least an essential stratum of that many-layered mystery. For if life is to be convincingly explained in terms of matter and energy, organization is all that stands between a soup of chemicals and a living cell.

From this point of view, a cell appears as a society of molecules, a dynamic pattern with dimensions spatial, temporal, and functional. The pattern is defined by its form and its functions, as displayed in a life cycle. Patterns reproduce themselves with high (but not perfect) fidelity and may persist in this manner for millions of years. The object of this chapter and the next is to examine what we have learned concerning the articulation of molecules into systems. How do millions, even billions, of molecules come to function in a collective, purposeful mode that extends over distances orders of magnitude larger? This, in essence, is the problem of biological order. We examine first the workings of a whole cell, and then turn to its growth and reproduction. Cells come, of course, in almost countless variety, and even among bacteria what is true of one is not necessarily true of all. But diversity must be set aside for the present; for clarity's sake I must focus here on a single organism, and that one can only be *E. coli*.

What's Spelled in the Genes

Simplicity, like beauty, is mostly in the eye of the beholder, and what the eye beholds depends on what it looks through. To the naked eye, *E. coli* is invisible; a suspension of 10 million cells per milliliter shows but a slight cloudiness. In the light microscope the cell appears as a short rod, about 2 micrometers long and 0.8 in diameter (the dimensions vary somewhat with strain and growth rate). No internal structure is to be seen, but with proper staining one can make out flagella and the nucleoid. The electron microscope reveals much more detail (Fig. 5.1), yet the adjective "simple" still applies. Thin sections display the cytoplasm packed with small grains, the ribosomes; a lighter central zone from which ribosomes are excluded, representing the nucleoid (two in the present instance, which depicts a cell beginning to divide); and a multi-layered envelope consisting of plasma membrane, the peptidoglycan cell wall and a lipopolysaccharide layer facing the exterior. Flagella, hair-like projections called fimbria and occasional inclusion bodies round out the list.

It is the biochemist who is privileged to behold the complexity concealed beneath that bland image (Table 5.1). There are more than 2 million protein molecules per cell, potentially of four thousand kinds, and nearly a thousand species of small molecules; 300 million molecules in all, not counting water which makes up nine tenths of the cell's mass. All these jostle one another in a volume of about one cubic micrometer,

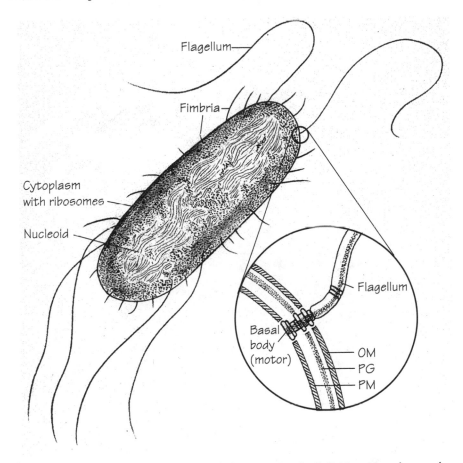

Fig. 5.1. The peerless cell: *Escherichia coli* at an early stage of cell division. Note the cytoplasm, densely studded with ribosomes, and the central exclusion zones that correspond to the two developing nucleoids. Flagella and fimbria emerge from the cell envelope. Inset: Cell envelope enlarged to illustrate plasma membrane (PM), the peptidoglycan layer (PG) and the outer lipopolysaccharide membrane (OM). Flagellar base enlarged to show the motor spanning the plasma membrane and the ancillary structures. Cell length, 3–4 μm.

a chamber far more crowded than the dilute solutions that biochemists prefer for laboratory studies (Fig. 5.2). Table 5.1 also features two strikingly small numbers: the cell wall peptidoglycan is covalently cross-linked, effectively turning the entire wall sacculus into a single huge molecule; and each nucleoid contains a single enormous molecule of DNA.

For the majority of scientists today DNA is the very essence of life; the god in the biological machine. Textbooks wax ecstatic over the master molecule that holds all the instructions required to make and

Table 5.1: Composition of E. coli*

Component	Molecules per Cell	Varieties of Molecules
Protein	2,400,000	1,850
mRNA	1,400	600
tRNA	200,000	60
(Ribosomes)	(20,000)	(1)
DNA	2.1	1
Lipid	22,000,000	50
Peptidoglycan	1	1
Small metabolites and ions	280,000,000	800
Water	40,000,000,000	1

*Data from Neidhardt and Umbarger, 1996, and elsewhere.

Fig. 5.2. A glimpse of the cytosol, magnified 1 million times. Ribosomes, proteins, and nucleic acid molecules jostle each other in this window 0.1 μm wide. From Goodsell, 1992, with permission of American Scientist.

run an organism. A quarter of a century ago, François Jacob celebrated that nucleic acid message, which records "the whole plan of growth, the whole series of operations to be carried out, the order and the site of synthesis and their coordination" (4); in fairness, let me add that subsequent pages sketch a highly sophisticated view of how the message is read and interpreted. Today, Richard Dawkins' influential writings (5) make the richest mine of evocative and provocative imagery. "Genes build bodies" to serve as their "vehicles" and "survival machines;" bodies are "robots," programmed by their DNA (here again, when fully spelled out Dawkins' views are more sober than they sound). Indeed, from an evolutionary perspective there is much to be said for regarding the history of organisms as an adventure of their genes, their DNA. Since genes are replicating entities, and the hereditary variations on which natural selection works occur almost exclusively in the genes, a DNA-centered view of evolution is illuminating, though at times a little shocking (I shall return to this thorny issue in Chapter 9). We are concerned here with quite another subject, the relationship between genes and bodies, genotype and phenotype, molecules and cells. Whatever the merits of phrases such as the genetic blueprint or recipe specifying the construction of a survival machine, they are all metaphors. To understand their meaning one must spell out what lies between linear sequences of nucleotides and the form and functions of cells orders of magnitude larger.

So compelling is the siren song of DNA that most scientists take it for granted that the genome specifies, not only the primary structures of proteins and RNA, but the higher levels of cellular organization as well. Scratch a colleague and you are likely to discover the conviction that a full understanding of gene regulation and genome structure are not only necessary for an account of organismic form and function, but virtually sufficient. *E. coli* allows us to examine this premise from an empirical standpoint, and the outcome suggests a less linear and more interactive appraisal of what the genes contribute to biological organization (6). Genes specify the cell's building blocks; they supply raw materials, help regulate their availability and grant the cell independence of its environment. But the higher levels of order, form and function, are not spelled out in the genome. They arise by the collective self-organization of genetically determined elements, effected by cellular mechanisms that remain poorly understood. If the genome is a kind of software, it takes for granted the existence of a particular and unique decoder. For myself, stubbornly determined to remain unwired, I prefer to think of the genome as akin to Herman Hesse's *Magister Ludi*: master of an intricate game of cues and responses, in which he is fully enmeshed and absorbed; a game that is shaped as much by its own internal rules as by the will of that masterful player.

*   *   *

The genome of *E. coli* consists of a single outsize circular molecule of DNA: 4.6 million base pairs, a molecular mass of 2.5 billion daltons (compared with 40,000 daltons for a typical protein), and a total length of 1,600 micrometers. Photomicrographs of the endless coil of DNA spilling out of a partially digested cell always remind me of the fisherman's genie, unfolding his enormity from the cramped bottle in which he had been immured. The nucleoid one sees in cell sections represents that DNA coiled, supercoiled, bundled and condensed into a minute volume (Fig. 5.3); how it remains accessible to the machinery of replication, transcription and translation boggles the imagination.

The circular gene map of *E. coli* has grown richer year by year for four decades; and just a few years ago the full sequence was published, completing a major stage in the determined effort to learn all about one

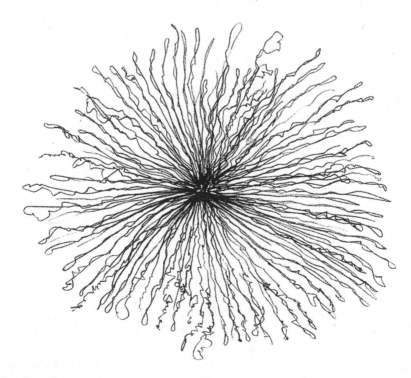

Fig. 5.3. An isolated nucleoid, after spreading the DNA on a protein monolayer. Scale bar, 1 μm. After Pettijohn, 1996, with permission of ASM Press.

species of life (7). The authors recognized 4,288 protein-coding genes, comprising 88 percent of the total genome; many of these proteins were known as biochemical entities, but nearly 40 percent were entirely new. Another 0.8 percent of the genome codes for stable RNAs, while 0.7 percent represents repeats that code for nothing. This leaves nearly 11 percent of the genome to accommodate regulatory extensions (binding sites for regulatory proteins, for example, or attenuators and enhancers of gene expression). These are not transcribed, but are obviously meaningful. All in all, well over 90 percent of the genome spells some useful function.

The remainder of the genome consists of genetic rags and patches: spacers that separate genes or gene clusters, non-coding repeats, transposable elements and cryptic bacteriophages and phage remnants. None of these are of obvious value to the cell, but they are nonetheless maintained for reasons that are generally obscure. The same applies to plasmids, circular DNA elements that inhabit the cytoplasm, replicate autonomously and are distributed between the two daughter cells at the time of division. Some integrate into the cell's chromosome, others do not. Plasmids often harbor beneficial genes (for example, genes that confer antibiotic resistance), but that is not necessarily the case and plasmids are better thought of as "selfish" elements that exploit the cell's metabolic machinery for their own reproduction. Whether or not they are overtly useful, plasmids confer upon the genome an unexpected degree of plasticity.

As genomes go that of *E. coli* is quite streamlined, presumably because a way of life governed by alternate bouts of feast and famine selects for compactness in all features so as to speed cell division. All the same, there is much apparent redundancy and even pointlessness: selfish plasmids, stretches that code for nothing, genes whose products duplicate the functions of other gene products. It seems unlikely that everything found on the great thread confers selective advantage, but neither is the genome just an artful contraption cobbled together from bits and fragments that happened to go well together. How much is chance, and how much design is one of the many deep questions we do not quite know how to ask.

A nucleotide sequence, like any text, can be seen as a linear string of symbols; it is of some interest to estimate how much information it can carry. Given that the selection of one base out of four requires two binary choices, any particular base (pair) can be assigned a value of two bits. The genome as a whole therefore bears a maximal information load of 9.2×10^6 bits or about 1 Mbyte, an amount of information that can be comfortably stored on a single floppy disk. In itself, this number is of limited significance. Remember that every alternative sequence of the same length, whether viable or not, boasts the same information content, that there is not a simple relationship between DNA content and other measures of biological complexity, and that a portion of even *E. coli's* genome seems to be devoid of functional meaning. Still, the bulk of it does represent meaningful instructions. For comparison, the information content of an English text can be estimated from the number of binary choices required per letter, about five. On that basis the information content of a volume of the Encyclopedia Britannica is pegged at 10^9 bits; and the *E. coli* genome is equivalent to about ten pages of that thousand page tome. The larger question is whether that thin sheaf contains all the instructions that a cell transmits to its progeny, and here the answer must surely be no.

How much genetic information does it take to make a cell? With the advent of whole-genome sequencing, a preliminary answer is already at hand. The first two genomes sequenced *in toto* were those of two small pathogenic bacteria: *Haemophilus influenza* with 1,703 genes and *Mycoplasma genitalium* with 469. These, of course, are actual organisms with specialized requirements and particular histories. Comparison of the two genomes, however, augmented with close reasoning allowed Mushegian and Koonin (8) to identify a subset of 256 shared genes that encode the essential functions. These include proteins that mediate DNA replication and repair; transcription, translation, and protein folding; enzymes for the metabolism of amino acids, nucleotides, lipids and cofactors; and catalysts involved in energy production, ion transport and secretion. Obviously, *E. coli* with a gene complement of thousands, has more than the minimal set, and the same will be true of any actual organism that makes a living in the real world. All the same, the number of indispensable functions seems remarkably small.

For the purpose of clarifying what the genes contribute to cellular order, it is essential to distinguish two levels of genetic meaning, the explicit and the implicit (9). As with any communication, the explicit meaning is that conveyed directly by the symbols of which the message is composed; the implicit meaning depends on the context of the communication, and embraces the ripple of consequences that proceed from the message. The phrase: "Can I help you?" may have any of a number of meanings, depending on the place, the circumstances and the tone of voice. In the present instance, the explicit meaning of a nucleotide sequence is the cognate chain of amino acids and, more remotely, the folded active protein. Note, however, that already the cellular context impinges upon the transfer of information, for the chain will only fold "correctly" in a medium of "appropriate" ionic composition and pH. Indeed, the very translation of the messenger RNA depends upon the cellular context, for ribosomes only work properly in the presence of high concentrations of potassium ions—a special milieu that the cell must do work to provide. Beyond the discrete stages of transcription and translation sprawls a web of structural and functional interactions that are nowhere spelled out in the genetic sequence but nevertheless make up an aspect of its meaning, at least as read by natural selection.

Monica Riley has taken upon herself the large task of keeping track of the gene products encoded by *E. coli* (10). At the present time we know something about 1,800 of these, less than half the total. Some gene products are known in detail, molecular structure and function as well as their place in the scheme of things; others are represented by little more than a map location and the phenotype of a mutant, but the tally is changing all the time. Three general conclusions emerge from this survey. First, a quarter of the known genes specify the machinery of transcription and translation; another quarter is concerned with the metabolism of small molecules, and one seventh with transport functions. These assignments make sense for an organism in which proteins make up half the cell's dry weight, and which responds quickly to changes in the nutritional circumstances by adjusting its complement of metabolic enzymes. Second, most, if not all, of the gene products perform quite mundane, nuts-and-bolts biochemical tasks: they are enzymes, or transport carriers, or species of tRNA. Third, neither the functions nor the structural organization of *E. coli* leap out from the table of gene products. These system properties must be sought in the even richer vein of meanings that is implicit in the genetic message but not spelled out there. The road from genes to cells leads through a hierarchy of epigenetic levels, upon which genetic information is progressively collated, integrated, and translated into the form and behavior of an organism. Its direction is defined, not by reduction and analysis, but by synthesis and growth.

THE DYNAMIC ORDER OF METABOLISM

What the economy is to human society, metabolism is to the cell. In short, metabolism is the ensemble of chemical processes by which cells obtain the goods and services required for their continuance, growth and reproduction. The metabolic web supplies energy and chemical building blocks, and it also carries the information that links the environment, the genome and the molecular machinery into a functional whole. More than three quarters of the known *E. coli* genes code for diverse enzymes and ancillary transport carriers. The web as a whole can be regarded as a device to generate complex structures and functions from simple starting materials plus energy. It is the metabolic web that converts energy into organization, building upon its own inherent orderliness.

Wall charts depicting cellular metabolism as a network of coupled chemical transformations have traditionally been a feature of laboratory décor; they display the channels through which matter flows in a segment of the metabolic economy, but omit the concurrent flows of energy and regulatory signals. The economic parallel is more than superficial. A cell is not unlike a city that supplies its residents with food, employment, education and medical care despite the virtual absence of centralized planning, and even copes with the bankruptcy of a major employer or a sudden flood. Both cell and city fall into the class of complex, adaptive systems. Adam Smith's invisible hand is a well-worn metaphor for the kind of order that is dynamic rather than structural, easy to recognize but hard to grasp.

In its early days, from the 1930s through the 1950s, the discipline of metabolic biochemistry developed quite separately from the abstract science of genetics. Research with bacteria quickly erased the division, for in these fast-growing cells the genome is closely integrated with all cellular operations. In *E. coli*, as much as half the gene complement is being expressed at any given time. The cells double in as little as 20 minutes, and adjust to changes in their environment by modulating their enzymic constitution on a time-scale of seconds to minutes. The genome specifies the molecular agents of the web, and gene expression in turn is regulated by products of the web. Microbial physiologists commonly refer to such regulatory circuits as "multigene networks," though, of course, proteins and small metabolites, together with the genes themselves, make up the working network. About a hundred such circuits are known at present, each a province of the global metabolic web, concerned with matters as diverse as ensuring the supply of phosphate and the constancy of cytoplasmic acidity, regulating the mix of respiratory enzymes, and coping with damage to DNA or with a sudden change in the availability of amino acids for protein synthesis. Let the last of these serve as an illustration of the dynamic, functional organization that is the hallmark of all of cellular metabolism.

When cells, grown on a rich medium supplemented with amino acids, are "downshifted" to a lean one in which they must manufacture amino acids for themselves, they face a major readjustment of the metabolic economy, and they mount an orchestrated, "programmed" set of responses that allows growth to resume with minimal delay. The immediate consequence of downshift is a shortfall of tRNA charged with amino acids, as consumption runs ahead of recharge; protein synthesis by the ribosomes is thus instantly curtailed. But a host of other responses quickly follows, affecting all biosynthetic activities. The production of stable RNAs (ribosomal and transfer RNAs) drops; the initiation of new DNA replication is halted; the rates of phospholipid and peptidoglycan production diminish sharply. These and other effects can be understood as part of an effort to adjust the manufacture of all cell constituents to the diminished capacity for protein production. At the same time, the biosynthesis of a subset of proteins—notably enzymes required for amino acid synthesis—accelerates markedly. Over time, as these enzymes progressively restore the supply of amino acids, restrictions on cellular operations are lifted and growth resumes (albeit at a diminished rate). This admirable and highly adaptive set of responses is elicited and coordinated through a small number of cytoplasmic signal molecules, modified guanosine nucleotides designated ppGpp and pppGpp, which are formed by idling ribosomes stalled by the lack of charged tRNA. The signal molecules, in turn, impinge on various regulatory loci, enhancing the activity of some genes and shutting down others. When the supply of amino acids is restored and ribosomal translation resumes, production of ppGpp is curtailed and the excess removed by degradation. The details, many of which remain to be worked out, are beyond our scope here; but I must at least mention that mutants deficient in the production of ppGpp fail to execute the foregoing reflex, and in consequence require many hours to recover from the downshift.

Is this pattern of purposeful chemistry laid out in the genome? Not in any identifiable form, and no one could have inferred its existence from a knowledge of the gene sequences or even of the gene products. What is encoded are all the elements of the pattern, complete with sequences that determine binding affinities and catalytic activities, regulatory sites and receptors, even localization signals that direct one protein to the membrane and another to the cytosol. Most of the proteins in question diffuse within the cytoplasm, colliding at random to generate interactive patterns that have functional significance but lack physical structure; their collective behavior should be increasingly amenable to computer modelling and simulation. Although the execution of the downshift-reflex is the task of proteins and other gene products, mutation and selection have shaped them so as to optimize the performance of the response as a whole. And since much of cellular physiology is subsumed under one multigene network or another, we can consider the entire metabolic economy to be implicit (albeit remotely) in the genetic instructions for the collective of its individual components—so long as we keep in mind the metaphoric nature of this usage. As matters stand, the nucleotide sequence of a gene is not sufficient to let us predict the three-dimensional structure of an enzyme, let alone its kinetic characteristics; the coordinated operation of a metabolic module is quite out of sight. Whether the metabolic economy can be "in principle" reduced to the molecular level is irrelevant to the question that demands an answer now. How does a coordinated, purposeful economic society emerge from the interactions among its multitudinous molecular citizens?

The molecular mechanisms that underlie the regulation of enzyme activity and gene transcription are generally understood, and some are known in exquisite detail. The unifying principle is allostery, the alteration of shape. The catalytic activity of an enzyme depends critically on the configuration of the enzyme protein, and may be enhanced or inhibited by any change in its morphology. Such alterations may be elicited by specific binding of an ion, a small metabolite or another protein; others result from chemical modification of the enzyme, its phosphorylation (or dephosphorylation) in particular. The ligands bind to sites remote from the catalytic one and play no part in the reaction mediated by the enzyme; regulatory molecules are pure symbols, as arbitrary as the letters of the alphabet, and display one more level of evolutionary design. Specific binding and conformational changes also underlie the activation and repression of genes, which are generally mediated by proteins that bind to particular sites on DNA. The theme comes with numerous variations, some of which are discussed in a thought-provoking article by Dennis Bray (11) that examines proteins as information-processing devices. In simple instances, molecular mechanisms suffice to explain physiology; more commonly, understanding must be drawn from the connectivity of the web, the pattern of interactions among numerous components. My object here is merely to point out some principles that give structure to the metabolic system.

The biochemist, looking at connections from the standpoint of molecular mechanisms, is impressed by their diversity. Unities are more conspicuous when one notes the technological parallels: pipes, valves and reservoirs; feedback regulation and feed-forward; the amplification of signals and their attenuation; tuners and switches; switches that respond to the convergence of two inputs, others that respond to either the one input or the other. It is the pattern of wiring rather than the molecular mechanisms that distinguishes one network form another.

The metabolic web can be dissected into modules, most of which perform some fairly discrete cellular task: degrading arabinose, regulating the osmotic pressure, coping with oxidative damage. Is this order inherent in the nature of things, or do our minds put it there, as when we see George Washington's portrait in a rugged rock face? The observation that mutations often disrupt a single physiological function while leaving others intact suggests that modules are real, but pleiotropic effects are no less common, and report the overlaps and links between modules. Modular organization also makes evolutionary sense, recalling the elaboration of the web by tinkering with its periphery rather than by rational redesign.

Modules are commonly arrayed in a hierarchical fashion. The genes that encode *E. coli's* metabolic enzymes are all members of one "operon" or another, contiguous genes that are transcribed and regulated more or less as a unit. But coordination reaches much higher. Neidhardt and Savageau (12), who grapple with the higher levels of hierarchical order and its diagrammatic representation, speak of regulons (a set of operons that share a common regulator; ppGpp is an example); modulons (a set of operons that are regulated individually as well as collectively); and stimulons (a set of operons that respond to a common environmental stimulus, but employ several regulators to do so; the response to a nutritional downshift, described above, may ultimately prove to be of this kind). One can readily construct intricate higher order circuits, and also recognize them in the global metabolic web. Note again that these functional modules are abstract, lacking physical structure. As far as we know, the interactions among the molecules in question are governed solely by their chemical specificities, encoded in the genes, and they encounter each other chiefly by diffusion and collision. One cannot help wondering whether that libertarian image is altogether correct.

Enzymic reactions are reversible in principle but unidirectional in practice, and the direction is ultimately imposed by energetics. The second law mandates that all processes in the real world flow down the thermodynamic hill: all entail the degradation of energy, the loss of potential and the production of entropy. Biosynthesis, control and communication all come under this requirement, and can be considered energy-dissipating processes. The dynamic order of the metabolic web, a state far removed from equilibrium, is maintained by the continuous flow of energy from some environmental source to its eventual sink, namely heat. Enzymes channel the flow of energy (and the accompanying flow of matter), but they do not create the current. The metabolic web illustrates the fundamental principle clearly enunciated by Harold Morowitz (13), that the flux of energy through a system organizes that system. And more than a few speculative thinkers suspect that the origin of metabolism and of life is to be sought in the great stream of energy that passes over the earth's surface.

The metabolic web is a complex system, whose properties as a whole are not always obvious from the properties of its parts and the rules that govern their interactions. One such emergent systemic property is the stability of the web, its capacity to continue performing its numerous functions while shifting abruptly from a rich medium to a lean one, or from one carbon source to another. One might expect this to depend on fine tuning of all the rate constants and enzyme concentrations, a delicate balance that would be readily upset by mutation. In fact, computer simulations suggest that many network functions are quite robust, in the sense that the output need not depend on the precise kinetic parameters of each link. Metabolic states can even be stable enough to be transmitted from one generation to the next, independently of the genes, representing a genuine case of the inheritance of acquired characteristics (14). Other features can be traced to the energetics of the web. Systems that are maintained in a state far from equilibrium by the flow of energy are apt to oscillate. Oscillations of metabolite levels have apparently not been observed in *E. coli*, but are well known in yeast.

Finally, note that all the known metabolic webs come enclosed in separate bodies bounded by lipid membranes; even the great natural cycles of nutrients and energy are made up of discrete units. Why are boundaries necessary? For reasons of principle and also of practicality. Only an enclosed pod can maintain a constant composition, and retain the high concentrations of diffusible intermediates required for high rates of reaction. Boundaries commonly serve to segregate incompatible agents: enzymes that hydrolyze phosphate esters are useful in nutrition but would perturb cellular metabolism, and are therefore confined to the cell's external surface. Coordination and coherence depend upon enclosure. And finally, natural selection can only come into play when it chooses among discrete metabolic units. Membranes are no less a requirement for life than are genes.

There is clearly still much to be learned about how the strands and knots of metabolism work together as a coordinated economy. The question is whether understanding can be attained by building on known principles of physics and mathematics, or calls for new paradigms—perhaps one of those additional laws of physics that Schrödinger mused on half a century ago. Thus Mae-Wan Ho, in the course of her efforts to elaborate a physics of organisms (15), felt compelled to invoke resonant energy transfer to account for the coherent workings of cellular molecules. I am not persuaded that experimental observations presently at hand demand this kind of extension into quantum physics, but would not dismiss the possibility out of hand. If indeed new principles are needed, that should be signalled by the persistent failure of efforts to simulate metabolism as a nonlinear dynamic network. That has not happened, but perhaps only because the effort is still at an early stage.

Metabolic networks, as we see them in *E. coli* and other bacteria, are products of evolutionary design. Mutation tinkers with individual genes, but selection judges the performance of organisms. What kind of metabolic structure would one expect to emerge from this optimization? An interesting article by Mittenthal et al. (16) approaches this question from the standpoint of "reverse engineering": by examining the structure and operations of a device, one can sometimes deduce what problem it was designed to solve. The metabolic web of *E. coli* appears to them as a dual entity, composed of two complementary nets. Each net is a "distributor": it allocates limited resources to perform diverse tasks, and therefore has the structure of a hub with spokes. The metabolic web is one distributor, that converts nutrients into a limited set of key metabolites. The second distributor is the macromolecular net, which polymerizes key metabolites into proteins and nucleic acids. Since the enzymes of the metabolic net are themselves a subset of the output from the macromolecular net, the biochemical web overall has the structure of a hierarchy of distributors. The authors argue that such a hierarchy of distributors is likely to evolve under selection for high performance using limited resources. It remains to be seen whether arguments from general principles can be carried further, to "predict" specific features of the metabolic patterns encountered in the microbial world.

The Limits of Self-Assembly

We turn now to the generation of spatial order, and the first step is easy for it springs directly from the molecular level. Its essence is the spontaneous association of macromolecules into specific supramolecular complexes, which comprise many of the cell's standard parts. What makes this test-tube exercise so significant is that, in principle, molecular self-assembly requires neither a source of energy nor of additional information; the structure and function of the complex are implicit in the nucleotide sequences that specify its subunits.

For example, take ribosomes. Each ribosome consists of two subunits; the smaller subunit contains 21 proteins arrayed on a scaffolding of RNA, the larger 31 proteins and two species of RNA. It is possible to dissociate each of the subunits into its constituent macromolecules, and to induce these to reassemble in the test tube into functional ribosomes. And this is but one instance of many. Most enzymes consist of two or more subunits that associate when mixed in a suitable buffer, and some of these join larger multienzyme complexes, permanent or transitory. The two strands of DNA, separated by gentle heating, reassociate precisely upon cooling to restore the original double helix. Flagella and fimbria self-assemble in the test tube. Self-assembly plays a large role in the formation of the flagellar motor (17). Both the plasma membrane and the lipopolysaccharide outer membrane display structures that arise spontaneously when their molecular constituents are dispersed in buffer, and the same may be true of the compaction of DNA into a nucleoid. The classic instances of self-assembly come, of course, from the world of viruses, including the bacteriophages that bite the back of *E. coli*.

Professional propriety demands a quibble at this point. It is well known that phage assembly *in vivo* requires the participation of scaffold proteins that do not end up in the final virus particle (18). During ribosome synthesis, the initial RNA transcripts must undergo tailoring by hydrolytic cleavage and methylation, which calls for cellular proteins that remain outside the ribosome. Even given the finished nucleic acids, ribosomal self-assembly proceeds to completion only if the mixture is gently warmed at a particular stage, the reason is not known, but the observation suggests that in the living cell constituents external to the ribosome intervene in its assembly. Other cases involve modulation of self-assembly by phosphorylation or by small ligands. Evidently, no cell part is truly an island. As a biological concept, self-assembly reminds one of the pious fictions that pepper our political discourse: it skims lightly over patches of thin ice, yet embodies an important partial truth.

Well then, is the cell as a whole a self-assembling structure, in the sense in which the term was used in the preceding paragraph? Would a mixture of cellular molecules, gently warmed in some buffer, reconstitute cells of *E. coli*? Surely not, and it is worthwhile to spell out why not. One reason is illustrated by the earlier quibble: assembly is never fully autonomous, but involves enzymes or regulatory molecules that link the organelle to the larger whole. But there are also three more fundamental reasons why a cell cannot be a self-assembling structure. First, some cellular components are not fashioned by self-assembly, particularly the peptidoglycan cell wall which resembles a woven fabric and must be enlarged by cutting and splicing. Second, many membrane proteins are oriented with respect to the membrane and catalyze vectorial reactions; this vector is not specified in the primary amino acid sequence, but is supplied by the cell. Third, certain processes occur at particular times and places, most notably the formation of a septum at the time of division. Localization on the cellular plane is not in the genes, but in the larger system. Cells do assemble themselves, but in quite another sense of the word: they grow.

Incidentally, it is instructive to reflect on the converse question: would cells survive if all their activities were shut down while their structural fabric is maintained? Indeed they do. Bacteria are routinely stored in the freeze-dried state for years on end, and a fraction of the population revives when restored to nutrient medium. The practice testifies to the primacy of structural organization in the preservation of life.

If large-scale order is not encoded in individual genes, might there be some sort of mapping from the arrangement of the genes on the chromosome to the spatial organization of the cell? Apparently not. True, many bacterial genes are arrayed in operons, but the genome displays surprisingly little physical order beyond that elementary level. Active operons need not even be located on the same DNA strand: some are on the Watson strand, some on Crick, and they are not all read with the same orientation. The higher levels of functional order, designated regulons or stimulons, are made up of operons scattered around the circular chromosome. At one time it was thought likely that the genetic map as a whole records two successive duplications of an ancestral genome, but the predicted order of map locations has not been found and the hypothesis has fallen out of favor. From genes to cells is a journey without maps.

Molecular self-assembly represents an essential principle of biological organization, the first stage on the road from molecules to cells. But molecular self-assembly does not suffice to account for cellular organization, and this failure is highly significant for two reasons. First, it rudely contradicts the unspoken assumption that nothing fundamental is lost when we grind cells into a homogenate, and therefore, that when we know all about the molecular parts we will automatically comprehend how cells are articulated and how they function. Second, the limits of self-assembly bear upon the meaning of the genome. The instructions spelled by the genes are local, not global. But growth and many other cellular operations depend upon energy requiring, directional processes; self-assembly in the cell is directed, in space and in time. It follows that only within the context of a particular cell, which supplies the requisite organizing power, is it valid to say that the genome directs the construction and operation of that cell. It is true but subtly misleading to envisage a cell as executing the instructions written down in its genome; better think of it as a spatially structured self-organizing system made up of gene-specified elements. Briefly, the genes specify What; the cell as a whole directs Where and When; and at the end of the day, it is the cell that usually supplies the best answer to the question Why.

LEGACY OF A REVOLUTION

The next stage on the road from molecules to cells begins, somewhat unexpectedly, in the discipline called bioenergetics. In the physical world, events are predictable from the second law of thermodynamics: water runs downhill, warm bodies cool, the sugar cube dissolves in my coffee, things fall apart. Living organisms, by contrast, are active: they climb the stairs, generate heat, accumulate nutrients and manufacture complex molecules. None of this violates the second law, for the work that organisms do is properly paid for by an external source of energy, just as a washing machine performs its task at the expense of power generated by the local power station. But the activities of cells and organisms do pose the problem of "energy coupling": How do cells capture the potential energy inherent in a ray of light or a pinch of sugar, and harness it to the performance of the many kinds of work required to flourish and multiply?

The first general answer to this question, formulated more than fifty years ago by Fritz Lipmann, was outlined in Chapter 4. To recapitulate, the immediate energy donors for the performance of work are adenosine triphosphate (ATP) and some related substances. ATP participates chemically in the processes that it promotes, undergoing breakdown to adenosine diphosphate (ADP) and inorganic phosphate (Pi). The function of the great highways of cellular metabolism—respiration, photosynthesis and fermentations—is to drive the energy-requiring resynthesis of ATP from ADP and Pi and thus sustain the energy economy. This clear and simple framework made bioenergetics comprehensible, and for many purposes it remains adequate today. It also focused attention on the next generation of problems, namely, the mechanisms by which respiration and photosynthesis supply organisms with ATP.

When I was a graduate student in the 1950s, and for a decade thereafter, this was one of the most perplexing issues in biochemistry. We knew that in bacteria and mitochondria electrons pass from reduced substrates to oxygen via a cascade of electron carriers called the respiratory chain. The overall sequence of this bucket-brigade had been established, and it was recognized that the respiratory chain is firmly associated with lipid membranes: the plasma membrane in bacteria, the homologous inner membrane in mitochondria. The pathway of electron transfer in photosynthesis, albeit not then known in detail, was clearly of the same general nature. We understood that the electron transfer chains supply the energy that supports the phosphorylation of ADP to ATP; we even knew the enzyme that catalyzes the latter reaction, the ATP synthase, likewise membrane-bound. The puzzle centered on the manner in which electron transfer and ATP synthase are linked. Biochemists, fresh from the triumphant elucidation of the mechanisms by which ATP is produced in several fermentative reactions, expected to find essentially similar chemical linkages in respiration and photosynthesis. For more than a decade, hopes would rise now and again, only to be dashed when the proposed chemical intermediates proved illusory. With hindsight, the fault is plain: not lack of technical skill, but a flawed conceptual framework. The coupling, we now understand, is fundamentally not chemical but electrical.

In 1961, a splendidly eccentric English scientist named Peter Mitchell proposed a radical alternative to the biochemical wisdom of the day. Mitchell's chemiosmotic hypothesis was truly a revolutionary notion in Thomas Kuhn's sense, entailing a change of paradigm—a transformation of the conceptual framework (19). Fig. 5.4a summarizes the basic principle as we see it today (20). The electron transport chain is not merely plastered onto the membrane (in the context of *E. coli*, the plasma membrane), but is so disposed within and across the membrane that, as electrons wend their way from substrate to oxygen, protons (hydrogen ions, bearing a positive charge) are translocated from the inner surface to the outer. The membrane forms a closed bubble or vesicle, and is relatively impermeable to protons. In consequence, proton translocation generates an imbalance of charges across the membrane, an electrical potential, interior negative. In time, a pH difference may also arise, interior alkaline. Protons at the external surface therefore find themselves at a higher electrochemical potential than those in the interior: they are subject to a "pull," derived from both the electrical potential and the pH gradient, which draws protons back across the membrane into the vesicle's interior. Mitchell referred to this as a "proton-motive force," by analogy to the electromotive force of a battery; today we prefer the term "proton potential." The membrane's lipid bilayer is not permeable to protons, but the ATP synthase spans the bilayer and provides a controlled passage across. The enzyme is so

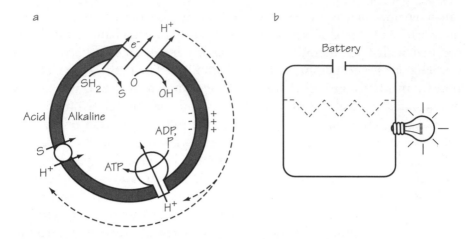

Fig. 5.4. Chemiosmotic energy coupling. (a) The general principle. Proton extrusion by the respiratory chain generates a proton potential, cytoplasm electronegative and alkaline. Protons return to the cytoplasm via the ATP synthase and by co-transport with substrate S. Arrowheads point in the physiological direction, but all the processes are intrinsically reversible. (b) The analogous electric circuit: a current of electrons generated by the battery lights the bulb. The dashed line illustrates how a conductive shunt can short-circuit the electrical coupling.

articulated as to couple the flow of protons "downhill" to the "uphill" production of ATP from ADP and Pi (The adjective "chemiosmotic" refers to enzymes, such as the ATP synthase, that catalyze simultaneously both a transport process and a chemical reaction). In Mitchell's view, then, the coupling of respiratory electron transfer to ATP synthesis is effected, not by a chemical interaction but by a circulation of protons across the membrane. What that proton current does is quite analogous to the role of the electron current in coupling a flashlight battery to the bulb (Fig. 5.4b).

Publication of the chemiosmotic hypothesis touched off a vigorous, sometimes acrimonious controversy over fundamental principles as well as experimental data; this continued for some fifteen years, subsiding only after the award of the 1978 Nobel Prize in chemistry to Peter Mitchell. By then the hypothesis had been as rigorously scrutinized as any proposition in biology, and judged to be essentially correct (21). The theory had also been broadened far beyond the coupling of electron transfer chains to ATP synthase. It was plain to Mitchell from the beginning that a proton circulation can support most kinds of work performed by membrane proteins. Imagine, for instance, a transport protein for substrate S (Fig. 5.4a). If this protein bears two functional sites, one for S and one for a proton, such that the flow of S is coupled to that of H⁺, then the driving force upon the proton is also applied to S. Such a transport protein will not merely facilitate the passage of S across the membrane, but act as a pump to accumulate S inside the vesicle.

A single cell of *E. coli*, bounded by its plasma membrane, corresponds in its entirety to the closed vesicle of chemiosmotic logic. (Fig. 5.5). When oxygen is present, the cells live by respiration. Protons are pumped outward by the respiratory chain embedded in the plasma membrane, generating a proton potential (cytoplasm negative). Protons return to the cytoplasm, completing the current loop, by any one of an array of routes, each of which harnesses the proton flux to some kind of useful work. The ATP synthase is one such: oxidative phosphorylation is the chief source of ATP for respiring cells, and ATP, in turn, supports cytoplasmic work functions such as protein synthesis or DNA replication. The plasma membrane also contains dozens of transport systems, each of which draws upon the proton current to pump specific nutrients into the cell or waste products out. Several enzymes use the current to do chemical work, including one that supplies reducing power for biosynthesis. Besides, as we shall discuss shortly, *E. coli* cells swim by means of rotating flagella; the rotary motor at the base of each flagellum is again powered by the proton current. Should the culture

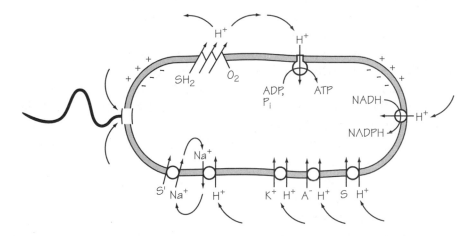

Fig. 5.5. E. coli, as the bioenergeticist imagines it. The diagram illustrates the generation of a proton current across the plasma membrane by respiratory proton extrusion. The current loop is completed by several paths that perform useful work: the ATP synthase (oxidative phosphorylation), the pyridine nucleotide transhydrogenase, the flagellar motor, and several kinds of proton-linked transport systems. Also shown is the generation of a secondary sodium current by Na⁺/H⁺ antiport and an example of sodium-linked substrate transport. S, substrate. After Harold, 1977, with permission of Annual Reviews, Inc.

turn anaerobic, *E. coli* can switch to an alternative energy source: the respiratory chain shuts down, but fermentable substrates supply ATP. Under these conditions the ATP synthase reverses direction: instead of using a proton current to produce ATP, it consumes ATP to pump protons out of the cell and maintain the proton circulation. In short, ATP and the proton circulation are alternative and complementary energy currencies, interconvertible via the ATP synthase. Here again we have a network of chemical and physical interactions, many of whose properties can only be appreciated by systems analysis and computer simulation. Just consider the many simultaneous processes that collectively determine the steady-state proton potential of a living cell!

It is not my purpose here to recount the experiments and arguments which eventually persuaded microbiologists that this is, in essence, how bacterial cells work; that has been done elsewhere (22). Let me instead highlight the transforming ideas that made the chemiosmotic hypothesis so controversial at first, and that now enlarge our perception of the cell as an integrated unit. First, many (perhaps all) enzymic reactions have an intrinsic direction in space. This is not apparent in solution, but becomes manifest when proteins are inlaid or plugged across a membrane; metabolic reactions then become vectorial. Second, linkage between two proteins need not be chemical, nor does it necessarily require direct contact; a flow of ions or metabolites can couple two vectorial reactions, as long as they share a common substrate and are embedded with the proper orientation in the same membrane. Third, coupling is a two-way street; either one of two coupled reactions can drive the other. Fourth, chemiosmotic coupling demands a topologically closed system such as a vesicle; a proton leak in either Fig. 5.4a or Fig. 5.5 would short-circuit the system. Finally, oxidative phosphorylation is an emergent property, and so are many other physiological activities. There is no gene for oxidative phosphorylation, though genes do encode all its elements. Only when a respiratory chain and ATP synthase are located in a single vesicle, and in the correct orientation, does the coupled reaction emerge.

Mitchell's hypothesis revitalized and transformed bioenergetics; I and many others built our professional careers upon it. But it has much wider significance. The chemiosmotic hypothesis explicitly introduced spatial direction into biochemistry at a time when the metaphor of the bag of enzymes was still prevalent (at least *in pectore*). It also taught us how nanometer-sized proteins could coalesce into micrometer-sized systems, and function coherently. Orchestration depends on membranes, on facing in the right direction and on being connected by currents.

Abstractly, the chemiosmotic theory illustrates the transition from molecules to cells.

READING THE MICROBIAL MIND

The human mind ranks high on every list of ultimate problems in science, and seems likely to remain there indefinitely. Perception, learning, memory, consciousness—when we probe our minds with our minds, confusion lurks 'round every corner. Bacteria have no room for the higher levels of mentation, but what minds they have can be read with remarkable clarity. The study of bacterial behavior is a fruitful and even noble endeavor, not because it holds clues to how humans think and act, but because it illuminates the essential nature of stimuli, perception and purposeful response. A bacterial cell is complex enough to register cues from the outside world and to respond in a goal-oriented manner; yet simple enough that its behavior can be dissected down to the molecular level and eventually resynthesized in the mind's eye in all its sophistication. The virtue of the exercise is to illustrate how a very intricate functional pattern is generated by a handful of catalytic proteins, communicating via diffusion constrained by localization.

Bacterial motility and behavior have drawn a particularly distinguished cadre of investigators, and one authentic hero: Julius Adler of the University of Wisconsin. Adler did not discover the ability of bacteria to detect sources of nutrients and swim in their direction; T. W. Engelmann and W. Pfeffer had done that in the 1880s. But Adler made chemotaxis a province of modern experimental biology. He identified attractants and repellants, introduced the use of mutations to analyze behavior, outlined the physiology and biochemistry of flagella and receptors, and first set the field in order (23). As I write this, thirty years later, Julius Adler is still actively engaged in the quest to understand how bacteria swim and search.

Cells of *E. coli* are propelled by their flagella (17), four to ten long slender filaments that project from random sites on the cell's surface. Each flagellum springs from a basal body that spans the plasma membrane and reaches into the cytosol. The flagellum passes through the cell envelope via a set of bushings and a "hook," and extends outward for several cell lengths; the filament proper comprises thousands of monomers of the protein flagellin, arrayed into a helical tube. Despite their appearance and name (from the Greek for whip), flagella do not lash; they rotate quite rigidly, not unlike a ship's propeller. By use of a tracking microscope devised for these experiments, Howard Berg showed that bacterial movements consist of short straight runs, each

lasting a second or less, punctuated by briefer episodes of random tumbling; each tumble reorients the cell and sets it off in a new direction.

The basal body contains the flagellar motor, a rotary engine driven
by the cell's proton circulation (Fig. 5.5), which can rotate the flagellum
either clockwise or counter-clockwise. Runs and tumbles correspond to
opposite senses of rotation. When the flagella turn counter-clockwise
(as seen by an observer situated behind the cell), the individual filaments
coalesce into a helical bundle that rotates as a unit and thrusts the cell
forward in a smooth straight run (Fig. 5.6). But not for long: frequently
and randomly the sense of rotation is abruptly reversed, the flagellar
bundle flies apart and the cell tumbles until the motor reverses once
again and counter-clockwise rotation resumes.

It is just this alternation of runs and tumbles that underlies bacterial
behavior. By tracking the swimming pattern of individual cells, Berg
found that cells moving up the gradient of an attractant towards its
source tumble less frequently than cells wandering in a homogenous
medium; while cells moving away from the source are more likely to
tumble. Each cell is engaged in a random walk rather than directional

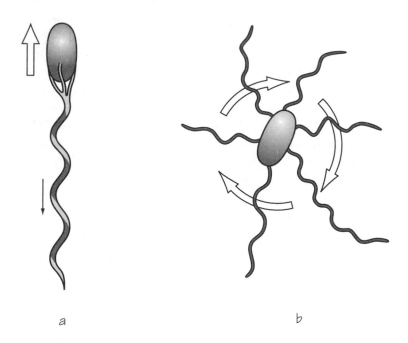

a   b

Fig. 5.6. Swimming and tumbling. (a) Swimming: the molecular structure of flagella is such
that when all the motors rotate counterclockwise, all the flagella coalesce into a single propulsive
bundle. (b) Tumbling: reversal of the sense of rotation causes the bundle to fly apart, and the
cell to tumble. After Macnab, 1996, with permission.

progress, but the attractant gradient imposes a bias that favors runs over tumbles. In consequence, the cell takes longer runs toward the source and shorter ones away, and rambles quite successfully up the gradient.

How can a cell "know" whether it is traveling up the gradient or down? In principle, it might "measure" the attractant concentration at its tip and its tail, and make the probability of switching dependent on the difference; for a cell only 2 micrometers long, the analytical precision required would be quite prohibitive. Instead, *E. coli* relies on temporal sensing: it measures the attractant concentration at the present instant, and "compares" it with that a few seconds ago. Since the cells swim at a speed of ten to twenty cell lengths per second, this reduces the precision required to one part in a hundred, still a very respectable achievement. Speaking metaphorically, the cell relies on short term memory to bias the probability that the motor will switch from counterclockwise rotation (run) to clockwise (tumble) at any given instant. A memory span of seconds may raise a smile, but it is quite adequate in the microbial world, where cells are continually buffeted by Brownian motion and can never hold any course for longer than a few seconds. And what these cells do is remarkable enough. *E. coli* can respond within milliseconds to local changes in concentration, and under optimal conditions readily detects a gradient as shallow as one part in a thousand over the length of a cell. Cells remain responsive to attractants over a concentration range of five orders of magnitude, nanomolar to millimolar. In effect, each cell performs a continuous series of rapid computations and acts upon their output.

The conceptual stage of this inquiry into the microbial mind was summarized in an important article by Daniel Koshland (24), in which he strove (in F. Jacob's phrase) to "give an account of what is observed by what is imagined." The initial step, measuring the attractant concentration, is the task of receptors situated at the cell surface which bind particular substances such as serine, aspartate and ribose. Signals from several classes of receptors, assessing a shifting spectrum of attractants and repellants, are transmitted across the plasma membrane into the cytosol. Somehow, the cell compares the "readings" of the present instant with those of a few seconds earlier, integrates them all into a unified signal, and conveys that to the flagellar motor. Koshland proposed that the probability of motor reversal is a function of the level of a hypothetical cytoplasmic substance, which he designated a "response regulator." The cell would contain a single pool of this substance, whose rate of synthesis and decomposition are variously influenced by different stimuli, and the time-dependent characteristics of this pool constitute the cell's "memory." Thus far, the imagination could be nurtured and disciplined by experiments with living cells: to proceed further it was necessary to take the cell apart and examine its working parts.

*   *   *

The object of contemporary research in chemotaxis is to account for the cell's behavior in terms of the structures and interactions of its constituent molecules. That goal has not been quite attained but is well within sight, and the focus of interest is beginning to shift from the molecular components to their integration into a sensory system. I cannot here do justice to the details of the circuitry; readers are referred to current reviews of the literature (25), whose essence has been distilled in Fig. 5.7. Let me then highlight how the parts fit together, making up an information-processing network that serves the cell's purpose. It takes about fifty genes to specify the proteins that collaborate in motility and behavior. Forty of these are required to construct the flagellar motor; we shall focus here on the remaining dozen which recognize stimuli, integrate the signals, transmit the output to the motor, and change its sense of rotation.

The first operation in the sequence, and the most sophisticated, is mediated by the chemoreceptor proteins plugged vectorially through the plasma membrane. There are five families of these in *E. coli*, each recognizing particular substrates: serine binds to the protein designated Tsr, aspartate to another called Tar, etc. The latter also recognizes mal-

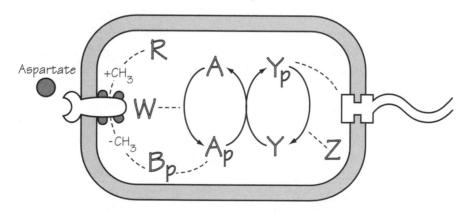

Fig. 5.7. Signal processing in chemotaxis. Binding of aspartate to the chemoreceptor protein initiates a cascade of conformational changes and phosphorylation reactions involving CheW, CheA, CheY and CheZ; flagellar rotation is governed by interaction of the motor with CheY$_p$. The response of the chemoreceptor is modulated by the degree of methylation, which depends on the enzymes CheB and CheR. See text for discussion. Diagram courtesy of Dr. Dennis Bray.

tose, but only in the form of a complex of maltose with a periplasmic binding protein. Association of the attractant with a chemoreceptor domain that projects from the external surface of the plasma membrane induces a conformational transition, which passes across the membrane and alters the activity of a second domain that projects into the cytosol. That cytosolic domain associates with two other proteins, designated CheW and CheA, and it is this complex that is responsible for the next stage. (The prefix Che refers to chemotaxis; these proteins were first identified as the products of genes in which mutations alter chemotactic behavior).

The CheA protein is a protein kinase; it catalyzes the transfer of a phosphoryl group to a particular histidine residue of its own, and subsequently transfers the phosphoryl group to the next recipient, the protein CheY. The rate of this phosphoryl transfer is a function of receptor occupancy. When the receptor is vacant the great majority of the cell's CheA proteins are phosphorylated, which ultimately favors tumbling. When the receptor binds an attractant molecule, CheA phosphorylation is inhibited, which ultimately promotes running. Binding of the attractant also promotes a chemical modification of the chemoreceptor, its methylation; we shall return to that step shortly.

CheY and its phosphorylated version CheY-P are crucial, for they correspond to Koshland's response regulator. As we presently understand matters, the CheY protein diffuses through the cytoplasm and carries the signal from the receptor-CheA complex to the flagellar motor. Binding of CheY-P to one of the constituent proteins of the basal body promotes clockwise rotation, and therefore tumbling; CheY does not bind, which allows the motor to turn counter-clockwise, and the cell to run. Thanks to the phosphoryl transfer cascade, the level of CheY-P is linked to receptor occupancy: vacant receptors produce CheY-P and induce tumbling, occupied ones favor the unphosphorylated CheY and hence running. Note that the phosphorylation status of CheY is also affected by another protein, CheZ, a phosphatase that reconverts CheY-P to CheY; just how this protein enters into the control of swimming behavior is unclear, but it may well play a major role.

Careful inspection of the signal transduction pathway sketched in Fig. 5.7 reveals a serious flaw. Each of the protein elements exists in the cell in thousands of copies (for example, there are thought to be some 12,000 CheY molecules per cell). When we speak of information transfer, we mean a shift in the average state of that population as a function of time. Suppose, now, that a tenth of the chemoreceptor proteins have bound aspartate. Since the signal elicited thereby is a negative one, and only a minority of the receptors are affected, the great majority of CheA and CheY molecules should continue to be phosphorylated, and therefore the motors should turn clockwise and the cells should tumble. In fact, the cells run and respond effectively to the gradient. What is missing is a channel that amplifies the signal generated by attractor binding. The nature of this channel is presently uncertain. Amplification may involve activation of the phosphatase CheZ; this should shift the balance from CheY-P to CheY, and thus promote counter-clockwise rotation and longer runs. Alternatively, the high "gain" may be a consequence of structural order, the clustering of chemoreceptor proteins into a much larger signal-transducing complex.

Where in this cascade of reactions and associations do we find the cell's memory? Not in any single step but in the state of the network that sets the fraction of CheY that is phosphorylated, and generates tumbling. One important input comes early, from the transfer of methyl groups to those chemoreceptor proteins that have bound an attractant molecule (the enzyme that mediates methylation is designated CheR). Methylation tends to reverse the signal conveyed by the attractant, but lags behind the phosphorylation of CheA and CheY; it should also trail the hypothetical activation of CheZ. This device allows the cell time to respond to one input before it is amended or superseded by the next. Methylation of the chemoreceptors (and their demethylation, catalyzed by CheB whose activity is regulated by the protein kinase CheA) serves a second function: adaptation on a timescale of minutes, analogous to our own eyes' capacity to adjust to the prevailing level of light. Adaptation allows the cell to ignore a constant input (high uniform serine) while remaining responsive to a second variable input (a gradient of aspartate, perhaps).

So far, so good; quite excellent, in fact. The molecular players have been identified, and each is being assigned a part. Moreover, the interactions between any two players can often be explained in terms of their chemical structures. The structural motifs that underlie receiver and transmitter functions; the precise nature of the conformational changes that receptors, CheY and the flagella themselves undergo; the mechanisms and degree of signal amplification and modulation at each step— these and other mechanistic questions now hold center stage. What remains to be done is to reconstruct the play: to work out how purposeful behavior emerges from the coupling between molecules whose individual characteristics are more or less known. The issue here turns, not on mechanisms but on relationships, and the appropriate technology is computer simulation.

It is quite possible to simulate the signalling pathway of chemotaxis in the simplified version illustrated in Fig. 5.7, with a single attractant and a single class of receptors, and even the early results are illuminating (26). Alon et al. found that the capacity of the network to adapt (reset to zero) in the presence of a constant stimulus is robust; that is to say, it does not necessarily depend on the precise concentrations and kinetic parameters of the elements, but can be a consequence of the network's connectivity. To make adaptation robust, the rates of methylation and demethylation (Fig. 5.7) must be modelled as functions of the phosphorylating activity of the complex, not of the concentration of the proteins. The qualitative features of the response to attractants fall out of the simulations (26), particularly when the stoichiometry of the various complexes is taken into account (receptor proteins, CheA and CheW are all thought to be dimers). The models also correctly predict the behavior of many mutants defective in the proteins, but some still elude the net, particularly those in which one or another of these proteins is present in excess.

Further progress may require a change in the conceptual framework. Dennis Bray (27) points out that the proteins which make up the signalling pathway are not free agents in a dilute buffer, but jammed together in a crowded cytoplasm that favors their association (Fig. 5.2). Signalling, it appears, reflects the operation of a structured and relatively permanent complex that includes the receptors, CheW, CheA and possibly CheZ. We may describe this body as a "signalosome," an integrated solid-state circuit that transmits information from the receptor to a mobile carrier, CheY. In such a cluster of chemotaxis proteins, a conformational change in one molecule may be propagated to its neighbor; this may be why the cells' swimming behavior responds so dramatically to a small increment in attractant concentration. A further, and quite unexpected, level of localization was revealed by the application of novel cytological techniques: signalosomes themselves cluster in discrete patches situated chiefly at the cell's poles (27). *E. coli*, it appears, has a "nose." What use it makes of its nose is a question for the future.

But we can already draw the conclusion that the purposeful chemistry of information processing takes place within the context of the cell's spatial order. The first step, and also the last, are vectorially disposed across the membrane, and signal transduction overall has a location in cellular space. Remember also that chemotaxis represents one channel in the cell's energy flux: signal transduction is a work function that consumes energy to maintain the steady-state levels of phosphorylated and methylated components, which is then modulated by the stimulus. Both of these considerations make it clear that chemotaxis is not just a matter of chemistry: bacterial signal transduction operates within that higher level of organization, which is superposed upon the molecular.

With the incorporation of the cellular perspective, a window opens upon a prospect in which all those signals and switches appear more as questions than as answers. The procedures employed by *E. coli* to sense gradients and respond appropriately are highly effective, but by no means universal among bacteria. *Bacillus subtilis* contains a protein homologous to CheY that causes the motor to turn clockwise, but here it is the unphosphorylated protein that signals tumbling, and the phosphorylated one that elicits runs. *Rhodobacter sphaeroides*, which swims by rotating its flagellum clockwise (rather than counterclockwise, as *E. coli* does), tumbles by stopping the motor rather than reversing it. Why did evolution shape different variations upon a common theme? Is each one specifically suited to the needs of its possessor, or does diversity notify us that the mechanistic details have no selective meaning? Such musings drive one back to the viewpoint of an earlier era, out of fashion in our day but perfectly valid for all that: "The organism in its totality is as essential to an explanation of its elements as its elements are to an explanation of the organism" (28).

How Structured is the Cytoplasm?

I don't suppose that anyone ever believed that a cell is just a bag of enzymes, but it makes a fine null hypothesis. Aside from visible organelles such as the nucleoid, the proposition goes, proteins and small metabolites diffuse without restraint wherever Brownian motion takes them. At the opposite extreme, one can imagine even a bacterial cell subdivided into compartments, its proteins anchored in place, with protein synthesis confined to one location and glycolysis to another. Reality falls somewhere between these poles, but just where is in question. Unlike eukaryotic cells with their conspicuous internal organelles and fibrous cytoskeleton, bacterial cytoplasm (and that of *E. coli* in particular) is featureless. Intracellular membranes are not normally present (even the nucleoid lacks a bounding membrane), and no permanent cytoskeleton has ever been documented; most of cellular biochemistry can be understood without recourse to order on the cellular scale. But the issue never quite goes away, perhaps because the image of millions of molecules colliding at random offends common sense.

Evidence for some constraints upon the ramblings of cytoplasmic molecules, particularly proteins, has been accumulating for years. The concentration of protein in the cytoplasm is as high as it is in some crystals, over 300g per liter of water; consequently the viscosity of cytoplasm is greater than that of water, and the rate of protein diffusion in cytoplasm is slower by as much as ten-fold. Besides, a substantial fraction of cellular water is loosely bound to protein surfaces; that fraction may be as much as a third or even half. These physical circumstances tend to favor labile yet specific associations between cellular proteins. Some metabolic enzymes occur in relatively stable multienzyme complexes, with metabolites passed on from one active site to the next without equilibrating with the bulk cytoplasm (29). It has even been claimed that all the enzymes of glycolysis are aggregated into a single labile complex nearly the size of a ribosome. Diffusion of membrane proteins is still further restricted; certain puzzling observations about oxidative phosphorylation suggest that, in natural membranes, respiratory chains and ATP synthases are closely apposed rather than widely scattered as Fig. 5.4 implies.

These observations, and others, underpin the proposal that the bacterial cytoplasm is integrated, not by a conventional cytoskeleton, but by the dynamic association among its proteins; Norris et al. (30) refer to this labile structure as an "enzoskeleton." Its elements may include fibers formed by proteins present in unexpectedly large amounts, particularly a soluble component of the machinery of protein synthesis designated EF-Tu that exhibits some similarity to actin. They also envisage the plasma membrane as a mosaic of small domains, differing in both lipid and protein composition: such domains will likely arise from the coupling of transcription, translation and insertion of membrane proteins (30). The organization of the nucleoid is not well understood, but if the loops of DNA prominent in Fig. 5.3 are permanent (or at least recurrent) features, then one can imagine genes as occupying relatively fixed topological positions; their products may then be directed to particular cellular locations. We are to envision both the functions and the structure of this enzoskeleton as subject to regulation, participating in such major localized processes as cell division.

At this stage of the inquiry, the only safe verdict is the Scottish not proven (or as they say north of the border, not guilty but don't do it again). But the evidence may shortly become much more persuasive, thanks to new methods of tagging proteins so as to make them visible to a microscope. That has already revealed the polar localization of chemotaxis proteins (preceding section), and of certain transport proteins. Proteins involved in cell division occupy defined cellular locations, and so do others that participate in the developmental cycle of visibly polarized bacteria such as *Caulobacter* and sporulating *Bacillus* (30). I expect to learn, in due course, that many bacterial proteins are in fact localized, perhaps not metabolic enzymes but most of those associated with DNA replication, cell separation and morphogenesis. The question then will no longer be whether the cytoplasm is structured, but how this order arises and persists; microbial physiologists will be compelled to look beyond the genes and rediscover the cell.

And Therefore, What?

This chapter set out to scout the swath of neglected territory that lies between molecules and cells, in the hope of catching a glimpse of principles that underlie biological organization. What can we conclude at this stage of our scrutiny of a single exemplar?

George Orwell once remarked that the first duty of an intelligent man is to state the obvious. So be it. A cell is an orderly society whose molecular citizens weave interactive patterns on many planes—spatial, temporal, functional, historical. Biochemistry is a science of molecules. Physiology deals in collectives. To put this more formally, a cell is a complex system (31). Its numerous and interdependent parts engage in accordance with fixed internal constraints, and in consequence the characteristics of the whole are invariant by comparison with the fluctuation of its constituents. Nature presents us with many kinds of systems, some simpler than cells and others vastly more complicated, but none that are more highly ordered (Chapter 2). The question here is how order (regularity) and organization (purpose, adaptation) come about in the molecular society that we designate a cell.

Several keys to understanding lie near to hand. A cell is made up of standard parts, arranged in a hierarchical manner that interact in ways governed by their molecular specificities. The cell is also a dynamic system formed by the confluence of several streams—flows of matter, energy, and information. It is spatially organized, not merely in the sense of having an external boundary, but in relying upon molecular processes that have location and direction. And it is endowed with a genetic program that ensures the accurate reproduction of all the working parts and is functionally tied into all cellular operations. The genome is also the seat of variations, whose systemic consequences are the substrate upon which natural selection acts. These system properties constrain evolution's freedom to experiment with novelty, since only mutations compatible with the existing pattern can pass the filter of natural selection. In this manner the phenotype feeds back upon the genotype, imposing a measure of stability and even direction upon change; tinkering

is possible, radical redesign is not. Evolution is thus directed into channels whose course can be traced in the homology of macromolecules
long after their functions have been transformed (see Chapter 9).

Bacteria are notoriously mutable, and might be expected to show
greater evolutionary flexibility than higher organisms. But here, too,
conservation is the dominant theme. *E. coli* and *Salmonella typhimu-
rium*, though closely related, are now thought to have diverged about
100 million years ago (32), say early in the Cretaceous! If not the dinosaurs, then the lumbering titanotheres of the Western fossil beds may
have harbored one or another of them. Biological patterns do change
over time as the geologist measures it, but not quickly.

The genetic program has rightly been the focus of intense scientific
scrutiny and of public celebration, but adulation has got out of hand.
The fallacy is the tacit assumption, taken as an article of faith, that *all*
the levels of biological order are spelled out in the genome. That is
obviously not true for *E. coli*, and *a fortiori* not for more elaborate cells
and organisms, without qualifying carefully just what is meant by such
throwaway phrases as "cellular functions are programmed by a genetic
network." First, it is clear that the cell (of which the genome is a part)
provides the context for the expression of that genetic network, mediates
its interaction with the environment and constrains its implementation
in space and in time. Second, many facets of a cell's complex and adaptive behavior arise from the interplay of its molecular components without the intervention of a central directing agency, just as the economy
of a city operates quite smoothly in the absence of plan or direction.
These system features, which come under the rubric of self-organization,
have been touched on several times in this chapter. They come to the
fore when we consider how cells make themselves during growth and
development, or divide and make two.

6

It Takes a Cell to Make a Cell

"The only way to get from genotype to phenotype is via development."

Scott F. Gilbert (1)

One, Two, Four and More
A Short Rod with Rounded Caps
Molecules into Cells: Paradigm Wanted

Rudolph Virchow (1821–1902) stood tall in the medicine and science of his day. He held the Chair of pathology in Berlin, wrote a major textbook and founded the Archiv für Pathologie, Anatomie und Physiologie, which he ran with an iron hand for more than fifty years. His specialty was cancer, and he was among the first to realize that cancer represents the malfunctioning, not of bodies or tissues, but of particular cells. But he is best remembered today for his succinct proclamation (1858) of one of biology's universal laws: *Omnis cellula e cellula*, (every cell comes from a preexistent cell).

Virchow's law has stood the test of more than 3 billion years, and with a little luck may outlast even the age of molecular biotechnology. Nor is it surprising, given what we now know, that it takes a cell to make a cell. Even those for whom life is simply the expression of the instructions encoded in the genes acknowledge that it takes cellular machinery to implement those instructions: enzymes, RNAs, energy, precursors, even the proper pH and ionic composition. But this can hardly be the whole story, for it fails to capture one of the key features of biological reproduction. Growth and division refer not simply to the accretion of biomolecules, but to the replication of an integrated pattern of functions and structures. These higher levels of order commonly depend upon molecular processes that have a particular direction or location in space. Reproduction is ultimately the business of cells, not of molecules, because direction and location are not spelled out in the genes; instead, a growing cell models itself upon itself. Michael Katz (2) has a word for it: the cell serves as the *templet* (not template), a source of configurational information, for the construction of its daughters.

We touch here upon another of biology's few general laws: Like begets like. It is true of *E. coli* as it is of elephants that offspring resemble their parents in form and appearance. How does that come about? The prevailing opinion is that form, like every other aspect of biological structure and function, is specified by the genes. There must be a large measure of truth in this phrase, for mutations commonly alter the shape of cells and even of higher organisms. We know instances in which the activation of a single gene switches the organism into a new developmental pathway, including the transformation of one form into another. And some basic features of differentiation in higher organisms, which hinges on the expression of one set of genes in muscle cells and another in liver cells, can truly be described as hardwired in the genome (3). On the other hand, genomes appear to contain nothing corresponding to the blueprint or recipe that is supposed to guide the physical construction of cells and organisms. In fact, the physiological procedures by which cells are built and rebuilt are only remotely connected with the genes (4, and Chapter 7). My interpretation of this apparent contradiction (shared with Katz and many other cellular physiologists past and present) is that every cell provides the templet upon which the daughter cell organizes itself. Many functions of that templet are performed at the level of genes and their immediate products, but the genetic instructions are supplemented by "cellular heredity", often carried in the cell's physical structure. Such heredity is commonly designated "epigenetic" (5), and the term templet is congruent with Conrad Waddington's celebrated image of the epigenetic landscape. If we are to understand just how like begets like, we must clearly specify the epigenetic mechanisms that create the visible shapes.

The proposition, that the continuity of cellular structure is a necessary complement to the continuity of genetic information, ought not to be seen as heretical. No one really doubts that the digital river of replicating DNA, springing from mysterious sources when the world was young, has always flowed within cellular banks. All the same, the focus on the cellular templet rather than the molecular gene represents a significant divergence from the genocentric conception of life that now dominates the scientific literature and even more so, the popular press. Professor

Doctor Virchow, for all his old-fashioned patriarchal mien, grasped a profound truth that is in danger of being forgotten in our own time.

The object of this chapter is to consider how one cell makes two, in the particular context of *E. coli*, and then to touch upon some of the broader implications of a view that makes the cell, rather than the gene, central to biological reproduction and morphogenesis.

ONE, TWO, FOUR AND MORE

Every cell's dream, François Jacob memorably observed, is to become two cells. *E. coli* is remarkably adept at living its dream, doubling in as little as twenty minutes. Indeed, making two cells where there was one before is the only discernible purpose of all the cell's metabolism, physiology and behavior. As life cycles go *E. coli's* is simple and straightforward, but it still consists of several parallel and concurrent tracks which together make up what is meant by the colloquial term "growth" (Fig. 6.1). One track takes in the production of cell constituents, which proceeds more or less continuously and everywhere so long as energy and nutrients are available and the conditions are favorable. Setting this aside, cell reproduction also calls for three processes that are closely localized in space or in time: (i) Doubling of the volume and surface area by elongation at constant diameter. (ii) Duplication of the genome and partitioning of the copies. (iii) Cytokinesis, division of the cytoplasm by the ingrowth of a septum at the midpoint, followed by separation of the daughter cells.

It goes without saying that the foregoing thumbnail sketch is grossly simplified. In particular, note that it takes no less than forty minutes to replicate the genome; cells manage nevertheless to divide every twenty minutes, by initiating new rounds of replication before the first one has finished. Consequently a fast-growing cell, such as the one depicted in Fig. 5.1, in fact, represents two cells about to become four. Many facets of cell structure vary with the growth conditions, including the cell's diameter and volume, and the presence or absence of flagella and other appendages, but I propose to set all these aside, so as to focus on the generation and reproduction of the basic architecture.

Bacterial cell division is presently an active research area, well covered from several points of view by recent literature surveys (6). The genetic approach has been especially fruitful, since even lethal mutations can often be studied in the form of temperature-sensitive alleles whose consequences are only expressed when the cells are grown at relatively high temperature. Normally the pathways of construction are closely coordinated, but many mutations are known to disrupt one or another of

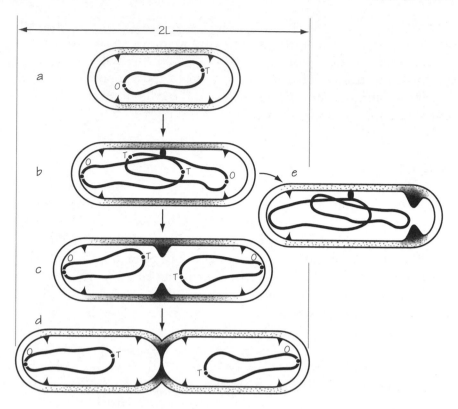

Fig. 6.1. One *E. coli* making two. The diagram emphasizes two aspects, wall synthesis and DNA replication. Stippling indicates the intensity of peptidoglycan synthesis. Symbols O and T mark the origin and terminus of replication, respectively. (a) A resting cell of length L. The spikes mark potential division sites left over from the previous cell cycle. (b) Growing cell. Sidewalls elongate by the dispersed insertion of new wall units, while the poles are inert. There is evidence that the DNA replicase is linked to the cell envelope near its midpoint; the duplicated strands attach to the poles. (c) Nucleoids separate; the cell assembles the septum precisely at the mid-point. (d) Septum closes creating two cells, each with one old pole and one new. (e) Aberrant division initiated at a silent site, producing a minicell without DNA. After several sources, none of whom should be held responsible.

them (Table 6.1). There are mutants that fail to form septa but segregate their nucleoids and grow into long filaments (designated *Fts* for fila-mentous temperature sensitive), others are defective in nucleoid segre-gation yet continue to elongate, and some cannot make a cylinder but grow into spheres with segregated nucleoids that eventually divide. These genes identify proteins that play essential roles in wall biosynthe-sis, cell division and morphogenesis, and the characterization of these proteins and their functions is a primary goal of contemporary research. Whether these thirty or forty genes should be said to encrypt cell form

Table 6.1: A Gallery of Division Genes*

Class	Gene	Phenotype	Biochemical Function
DNA Partition	*par* (several)	DNA clumped in cell center, growth and septation continue	Enzymes of DNA synthesis, gyrase or topoisomerase
	mukB	Similar to *par*	Possible motor protein involved in DNA segregation
Cytokinesis	*ftsA*	Filamentous at re-strictive temperature. Septa fail to form	suspected ATPase
	ftsI	"	Penicillin-binding protein 3, a peptidoglycan-modifying enzyme
	ftsQ	"	Unknown
	ftsZ	"	Ring protein, mediates septation
	envA	Septa do not split	Unknown
Shape	*pbpA*	Cells do not elongate, become spherical	Penicillin-binding protein 2, a peptidoglycan-modifying enzyme
	rodA	"	Unknown, acts in concert with penicillin-binding protein 2

*Data from Donachie, 1993, and Lutkenhaus and Mukherju, 1996.

and its reproduction seems to me much more questionable. Our object here is not to identify the molecular players but to comprehend the game; one seeks a causal and dynamic account of the process of cell reproduction, and this must turn on the timing of events and their localization in space. This goal has not yet been achieved, which makes it all the more important to underscore the central issues.

* * *

In bacteria, as in other walled cells, morphogenesis revolves around wall construction. Bacteria lack an internal skeleton, and it is the rigid cell wall that confers and maintains the cell's shape. We know this from the observations that wall fragments retain the shape of the cell from which they come, and that careful treatment of cells with enzymes that digest the wall yields spheroidal cells, dubbed spheroplasts. The crucial component for shape generation is the peptidoglycan layer sandwiched between the plasma membrane and the outer lipopolysaccharide membrane. Peptidoglycan has been likened to a fishnet, made up of polysaccharide chains crosslinked by short polypeptide strands into a continuous fabric. In effect, the entire cell is encased in a single huge sack-shaped molecule. The net is only one or two molecules thick and fully stretched; think of peptidoglycan as a strong, stiff but open-meshed fabric, not unlike nylon. The net is rather irregular; but in general incomplete polysaccharide hoops run around the cell's circumference, while the polypeptide cross-links run lengthwise (7).

What determines the shape of this giant macromolecule? It does not seem to be a matter of chemistry as such; despite much effort, no chemical difference between the peptidoglycan of poles and sidewalls has been discovered. Nor is the peptidoglycan a self-assembling structure: its subunits are covalently cross-linked, and the net can only be enlarged by cutting into the existing fabric and inserting a patch. Doubling the sacculus is no small matter: Park (7) estimates that each cell requires some 600,000 cleavage steps, insertion of over 2 million new units, and another million for the septum that ends up as a new cell pole. So the changing shape of a cell as it elongates and divides must be determined by where, when, and how the new wall is laid down.

Just to make it harder, the cell must maintain the wall's structural integrity at all times. A growing cell is a vessel under pressure—about 5 atmospheres in the case of *E. coli*, comparable to the air pressure in a racing bicycle tire. The reason is that the concentration of solutes in the cytoplasm is generally much higher than that in the medium, so water tends to flow into the cell. The wall is stretched but resists and constrains the influx of water, generating the hydrostatic pressure called turgor. Turgor is a problem for every walled cell: if wall enlargement weakens the wall so that it ruptures, the cell dies. But turgor is also part of the solution: most walled cells can grow only so long as they are turgid, and they possess elaborate regulatory mechanisms to maintain turgor. Turgor, it appears, supplies the driving force for wall expansion.

In contrast to the cell wall, the plasma membrane is a fluid bubble that accommodates itself to whatever shape is imposed by wall growth. Excess membrane, which is produced by certain mutants, accumulates in the cytoplasm in the form of blebs, vesicles and tubules. Phospolipids are synthesized by enzymes associated with the cytosolic surface of the plasma membrane, and are incorporated into the bilayer's inner leaflet; they diffuse into the outer leaflet via specialized junctions where the two leaflets merge. Thus far, membrane growth can be regarded as a kind of molecular self-assembly, but there is much more to it than that. To begin with, membranes do not arise *de novo*; phospholipids insert themselves only into existing membranes. Next, unlike artificial phospholipid bilayer vesicles (liposomes, Fig. 4.2), in which the composition of the two leaflets is the same, *E. coli* membranes are asymmetrical: certain phospholipids are found preferentially in the outer leaflet, others in the inner. Likewise, the insertion of membrane proteins is partly ordained by their primary structure, which specifies the hydrophobic segments of the folded protein that span the lipid bilayer. Insertion of the nascent membrane protein, however, requires both ATP and a proton potential of the correct polarity: self-assembly is modulated, even directed, by cellular influences.

Much the same is true of the outermost, or lipopolysaccharide membrane. Here again the lipid and protein components are produced at or by the plasma membrane, and travel to the exterior surface via specialized adhesion zones where the two membranes communicate. The basic lipopolysaccharide structure is a repeating one that can arise in the test tube by self-assembly. In the living cell, however, transport of the components to the outside depends on the proton potential, and their assembly takes place on a scaffold supplied by the peptidoglycan wall. I shall say no more about the generation of these two membranes, since they do not directly influence the form and division of the growing cell. But they, like the wall, illustrate the production of a spatially extended, multimolecular structure that cannot be understood solely as the product of individual biosynthetic enzymes but reflects the integrated workings of the cell as a whole.

Finally, a word about the nucleoid. As someone once said in reference to the chromosomes of eukaryotic cells, the nucleoid's role in division is like that of the corpse at a funeral: it is the reason for the proceedings but takes no active part in them. Unlike eukaryotic cells, bacteria lack a mitotic spindle and, indeed, a true cytoskeleton; instead, the segregation of the nucleoids after DNA replication is closely coupled to the extension of the cell envelope. This idea was first formulated thirty years ago by Jacob et al. (8), who imagined the DNA to be linked to the membrane; and while the specifics require renovation, the essential conception persists in contemporary hypotheses.

A SHORT ROD WITH ROUNDED CAPS

How forms arise and reproduce themselves generation upon generation is not clearly understood for any cell or organism, but in the case of bacteria we have at least a coherent framework of ideas that can be applied to particular cases. The surface stress theory is the brainchild of a single individual, Arthur Koch, of Indiana University; and while Koch has always been at pains to acknowledge the input of collaborators and critics, he is clearly the one who (to borrow a phrase from Freya Stark) found the facts in a tangle and sorted them out in a pattern of his own. Surface stress theory has been evolving for nearly two decades now (9); it anticipated many of the experimental findings of recent years, weaves them into a rational pattern, and is undergoing continuous revision. The theory is far from universally acknowledged, and receives only passing mention in *E. coli's* Good Book (10). But it presently offers our best hope of understanding the essential nature of bacterial forms, and it is quite remarkable that we have no alternative answer to offer anyone who asks how bacteria grow and shape themselves.

The core of the surface stress theory is the identification of the forces and constraints that shape bacterial cells. Unlike animal cells, bacteria lack a cytoskeleton and associated mechanoproteins; their form is molded by the biophysics of the cell envelope, and the relevant force is turgor pressure (11). As cells grow and accumulate metabolites, turgor pressure stretches and stresses the wall in patterns that depend on its geometry. The cells respond by inserting new peptidoglycan units into the existing wall, allowing it to expand. The general principle is global force, local compliance. The resulting shapes—cocci, rods, filaments, etc.—reflect the patterns of localized wall expansion, and they are produced by internal hydrostatic pressure acting on the plastic zone of the wall as it expands. Concurrently, turgor pressure performs another role: it relays information about the cell's metabolic state to its envelope, and thus couples the cell's economic success to volume enlargement and ultimately to division.

According to this way of thinking, bacterial growth bears some analogy to the behavior of soap bubbles. Setting aside for a moment the obvious differences, what they share is the physics: a fluid envelope, expansion driven by pressure, the smallest surface area compatible with surface tension and mechanical constraints. The surface stress theory grew out of Koch's serendipitous realization that streptococcal cell poles assume the same shape as fused soap bubbles, and the parallels have been drawn in some detail. Contrary to common belief, soap bubbles need not be spherical; given the proper support and relationship between pressure and surface tension, even a cylindrical bubble can elongate in a stable manner. Moreover, while the solid cell wall appears quite unlike a fluid soapy film, at the instant of expansion one can treat the wall as though it were fluid and describe its behavior by an equation derived from the physics of bubbles. This has allowed Koch and his collaborators, in a few favorable instances, to calculate the shape a cell should have (9). On the other hand, the analogy should not be pushed too hard: unlike soap bubbles, whose form is wholly predictable from physical forces and mechanical constraints, bacterial cells exert some physiological choice in the matter.

The biological fulcrum comes at the point of localizing wall expansion, this is effected by localized enzymes positioned by spatial markers, and is thus linked both to the genome and to cellular architecture. Once a new wall unit is exposed to hydrostatic pressure, physics determines the shape. Consider, for example, cell division in enterococci (formerly streptococci), perhaps the simplest instance of bacterial morphogenesis (Fig. 6.2a). An enterococcal cell has the shape of a football, with a prominent wall band girdling its circumference. When growth begins, the wall band splits lengthwise; new peptidoglycan is mostly laid down

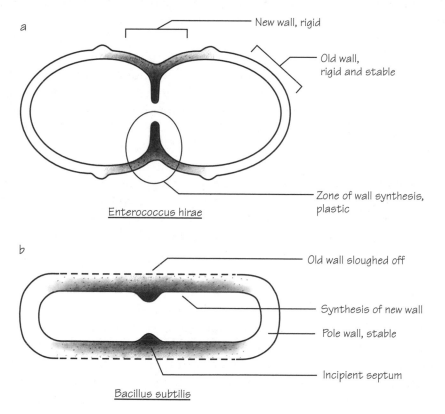

Fig. 6.2. Bacterial shapes result from the interplay of localized wall synthesis, turgor pressure and surface stress. For explanation see text. Stippling indicates the intensity of peptidoglycan synthesis.

in the zone between the two halves, and division concludes with the construction of a septum across the cell's midriff. As the cell's volume increases, so must its surface area. The growing septum splits and progressively becomes part of the external surface; as septal material comes under tension, turgor pressure distorts what had been a flat partition into a dome. An individual enterococcus thus consists of two poles, one old and one new.

Bacillus subtilis, a short rod, employs a different strategy (Fig. 6.2b). Hydrostatic pressure can support and enlarge a cylinder, so long as certain constraints are met. Those include the requirement that sidewalls and poles be made alternately, not concurrently. Sidewalls must extend by the diffuse insertion of new wall units (as they are now known to do), while the poles provide rigid support. In these Gram-positive bacteria the wall is many layers thick, and there is no outer membrane. Consequently, as new wall units are inserted into the innermost layer, the outer ones are sloughed off ("inside-out growth"). Poles, by contrast, are only produced when a dividing cell lays down a septum; once made, they remain stable.

A bacterial cell is a vessel under pressure, and employs well-known engineering tricks to cut and splice the wall peptidoglycan without risking rupture (9). One is to insert new units under stress-free conditions. A case in point is the formation of a septum by ingrowth of the envelope; only when the septum has been closed, split, and externalized as a new pole does it come to bear stress. The other trick is the hypothetical principle that Koch calls "make-before-break": hydrolytic enzymes must not cleave bonds in the peptidoglycan fabric at random, but should recognize and sever only bonds protected by a new wall unit attached and ready to slip into place.

Well then, how would *E. coli* shape a form that can be idealized as a cylinder with hemispherical poles? The basic presumption is that *E. coli* grows as a cylindrical soap bubble, much like *B. subtilis* (Fig. 6.2b). Extension of the sidewalls with constant diameter is compatible with surface stress physics, on three conditions: the insertion of new peptidoglycan units must be dispersed along the cell's length (rather than being confined to a localized growth zone), the cell poles must provide rigid and stable support, and the ratio of turgor pressure to a parameter corresponding to surface tension must fall in the right range. Recent experiments confirm that the first two of these requirements are indeed met. Cell poles are laid down as septa, and once completed remain stable; sidewalls extend by random insertion of new units at about 200 sites per cell. Tension within the wall is probably responsible for the arrangement of peptidoglycan chains into loops around the circumference, while the cross-links run lengthwise. There is in principle no need to invoke hoops or stays to account for the constant diameter of normal cells, but one wonders whether soap-bubble physics is sufficient to account for the long aseptate filaments of certain mutants. Few investigators would be surprised to learn that mechanical reinforcement is part of the design (12).

Physics grants *E. coli* license to extend as a cylinder, but does not explain how the trick is done; biochemistry is needed for that, and especially to explain how a thin net stretched to the limit can be safely cut and patched. The general shape of *E. coli* resembles that of *B. subtilis*, but where the latter features a thick wall that can securely elongate by inside-out growth (Fig. 6.2b), the peptidoglycan layer of *E. coli* is only one or two molecules deep. A clue to the strategy employed (one version of "make before break") comes from a curious feature discovered a decade ago: as the cells grow, sidewall peptidoglycan is simultaneously elongated and broken down to recyclable fragments. This finding, together with his discovery of a novel triple linkage in peptidoglycan walls, led J. V. Höltje to formulate an explicit chemical mechanism for safe wall expansion. It calls for a clever multifunctional enzyme that links three new glycan strands to the existing fabric and then cuts out one of the old strands, letting the new patch slip into place ("three-for-one"; 13).

When the cell has grown to the "right" size, division begins. The nucleoids, whose replication has been under way for some time, separate and condense in their respective cell-halves. Midway between them, the plasma membrane invaginates, followed by the envelope, and the septum begins to grow into the cytoplasm (Fig. 6.1). In due course, the septum closes, splits lengthwise and bulges out into a pole, the daughter cells go their separate ways, each consisting of a new pole; an old pole and a recycled cylindrical sidewall. A slew of questions hangs by this simple tale, with partial answers at best. How does the cell know when it is the right size? Perhaps by measuring a parameter that reflects its mass or volume (14). And how are the replicated DNA genomes separated? It now appears that they become linked to the cell poles by the origin of replication, and are separated as the sidewalls lengthen, but there are recent indications that they also receive a sharp shove at the end (15). The most baffling aspect of cytokinesis is the growing cell's ability to locate its mid-point with high precision. Of the several hypotheses that have been proposed (16), I shall sketch only one that fits with the biophysical tenor of this section. Koch and Höltje argue that, thanks to the composite structure of the envelope, tension within the plasma membrane of growing cells will be maximal at the mid-point. They postulate a sensor in the membrane which is activated at the locus of maximal tension and initiates septum ingrowth.

Ingrowth of a septum can be expected on biophysical grounds, but the research spotlight is presently focused on the proteins that actually produce the new wall. The ferment was touched off by the discovery that one of these, FtsZ, is assembled into a ring at the site of the incipient septum (the Z-ring), which guides its progressive closure (17). The details, which are rapidly being worked out, need not detain us. Note, however, that if the emerging conception proves true, we have here a striking instance of molecular self-assembly localized and directed by cellular physiology in both time and space. As a matter of fact, there are two cellular procedures (employing a common set of proteins) to establish the site of septum ingrowth. In one, sketched above, the cell finds its midpoint in a manner yet to be established. In the other, it may re-use a site left over from a previous cell cycle, which now forms the junction between sidewall and pole (Fig. 6.1a). *E. coli* does not normally make use of that site, but certain mutants do, cutting off minicells devoid of nuclei (Fig. 6.1e). This phenomenon illustrates another general principle: the use of positional markers to localize physiological events.

Slowly, painfully, we are muddling towards an answer to the question how *E. coli* grows a short rod with rounded ends. But I have never found a persuasive answer to the question *why* it should do so. What selective advantage accrues to *E. coli* from being rod-shaped? It is, as Donachie and his colleagues point out (14), an efficient shape for segregating genomes, but the world is full of cocci and stranger shapes that manage just as well. The same caveat applies to the notion that elongated shapes have a more favorable ratio of surface area to volume. Could it be that the elongated shape is hydrodynamically more efficient, and therefore advantageous for motile cells? More generally, does form necessarily follow function? Could it represent the expression of a higher-order morphogenetic law, or simply the unselected outcome of *E. coli's* particular physiology? If you know the answer, or have a well-grounded opinion, do please let me know.

MOLECULES INTO CELLS: PARADIGM WANTED

I seldom open the hood of my car, other than to check the oil, for the tangle of wires, tubes, and terminals is unintelligible to me. I must take my mechanic's word for it that they make the wheels go around and ensure my safety, comfort and compliance with the law. The massive compendium of *E. coli* lore by Neidhardt et al. (1996) stirs similar feelings: Most of it is true and much of it is significant, but the sheer mass of information overwhelms the faculties. Unfortunately, in this instance there is no higher authority to which to turn. There is also no firmer ground on which to reassemble the whole cell from its multitude of molecular components, at least in the mind's eye.

The prevailing framework for thinking about biological organization takes the form of the genetic paradigm. It builds upon the established relationship between genes and proteins but vastly enlarges its scope. Broadly, the thesis is that a cell's molecular composition, structural anatomy, form, and behavior are all determined by its complement of genes. That need not mean that all these matters are spelled out in the genes, nor that a cell can be physically mapped upon its genome, but rather that the genome constitutes something like a recipe for producing that cell. Stated thus, the genetic paradigm is not a hypothesis drawn from testable propositions, but what is called in German a *Weltanschauung* (a world view) that expresses the reductionist spirit of contemporary science.

Growth and morphogenesis in *E. coli*, the classical proving ground of the genetic paradigm, force one to ponder the inward meaning of that potent metaphor. Form and development clearly come under the genes' writ, as attested by the many mutants in which these features are altered (Table 6.1). Though not explicitly specified in the genetic instructions, a rod with rounded caps must be one implication of the set as a whole; just as the delicate flavor of the marble cake that my wife bakes on special occasions is nowhere mentioned in her notes. But wait. The products of those genes, insofar as we know them, turn out to be enzymes and other proteins that perform essential but mundane and local tasks. Knowledge of the genes and what they encode is nowhere near sufficient to explain how the cell elongates, divides and shortly produces a pair of rods with rounded caps. What we seek to understand emerges from the sociology of molecules, not their chemistry, and carries us into a different layer of reality. Indeed, how could it be otherwise? A growing cell is not a self-assembling set of puzzle pieces, but the product of generative processes mediated by multiple molecules, physiological pathways deployed in space. The reactions that shape a cell have, of course, a chemical dimension; but unlike their fellows in the test tube, many of them display direction, location and timing. Cell biology is about chemical and physical events that take place here rather than there, transport matter from here to there, not now but later, when called for. Once your eyes have been opened to these upper levels of order (as mine were by Peter Mitchell thirty years ago) you see them everywhere. *E. coli*, like all cells, practices biochemistry with an attitude:

Vectorial transmembrane reactions;
Vectorial insertion of membrane proteins;
Localization of chemoreceptor proteins;
Directed assembly of the FtsZ ring;
Diffuse elongation of sidewalls;
Finding the cell's midpoint;
Spatial markers for morphogenesis;
Orientation of murein links by stress;
And many more.

So cellular organization is chemical and molecular, bred in the genes; but a cell reaches much farther, flaunting capacities that are rooted in the operations of the larger unit. If you like to think of the genome as software, then cellular organization corresponds to the interpretation of the program by its own unique reader (18). This way of thinking gives one a more realistic appreciation of that peculiarly biological kind of self-assembly commonly known as "growth." Perhaps if growth were not part of everyday experience, we would see how wondrous it really is. DNA makes RNA makes protein is one of the universals of (terrestrial) biology. But what common principle underlies the dependence of protein folding on the ionic environment, the emergence of a capacity for ATP synthesis when several proton-translocating pathways come together in a single vesicle, the localization of chemoreceptors to a cell pole, and the directed assembly of the FtsZ ring at the cell's midpoint? None, except that they all obey both the digital dictates of a stretch of DNA and the subtler promptings of the epigenetic landscape. That term, coined by Conrad Waddington more than forty years ago (5), helped him visualize the forces and constraints that guide the trajectory of a developing embryo. It remains an indispensable metaphor for those of us who strive to identify and make intelligible the levels of order that intervene between molecules and cells.

Every particular aspect of growth and development poses questions of mechanism, with answers to be formulated in terms of molecules and the forces that act upon them. And with every new issue of the journals we can make out more clearly what form the answers will take, even though the specifics remain to be settled. Genes can be activated, or silenced, by the methylation of DNA or the judicious placement of transcription factors. We already understand in general terms how membrane proteins are inserted with the correct orientation, and how vectorial chemistry comes about. Earlier in this chapter, I sketched speculative proposals by A. L. Koch and J. V. Höltje for wall enlargement and finding the cell's midpoint. Quite recently, Shapiro and Losick (19) have reviewed, in exemplary fashion, what is known of the localization of proteins in bacterial cells. Now that methods are being developed to determine which protein goes where, it will not be long before we work out how each is delivered to its destination. These, to my mind, are puzzles. The mystery lingers just beyond. What pulls together the cacophany of molecules and ion channels and regulated pathways into a coherent whole: a cylinder with rounded caps, quickly and every time? If a cell is an orchestra and DNA the score, who or what conducts?

My colleagues in molecular science seem determined to ascribe that role, also, to the genome; for reasons given above, I find that attribution unbelievable. A rather more plausible candidate is the plasma membrane, which defines the cell. The membrane creates the enclosed and controlled space, within which societal behavior emerges from the interactions among individual molecules. Molecules are commonly oriented with respect to a membrane. Enzymes that share a membrane, or respond to a common signal, operate coherently; thus far, I see no need to invoke coherent excitation, a phenomenon of quantum physics, as Mae-Wan Ho has done. Developmental processes commonly begin at the cell surface and proceed inward; assembly of the FtsZ ring is the pertinent example here. If the bacterial plasma membrane represents a mosaic of domains that differ in composition and function, as various recent findings suggest (20, and Chapter 5), then the plasma membrane plays a still larger role in structuring cellular space. But the organizing powers of the plasma membrane, and of a transient cytoskeleton, are obviously limited. They may do for concertmaster, but not for conductor.

Here we reach an edge, and are left contemplating the disquieting notion of an orchestra without a conductor. Physicists use the term self-organization to describe what happens in a system whose constituents convert from individual, random motions to a state of global cooperativity; a favorite example is the Bènard instability, the emergence of a pattern of convection cells in a pan of oil uniformly heated from beneath (Chapter 7). If we are prepared to tolerate some creative ambiguity, self-organization will also serve to underscore the distinctive feature of the biological systems: spatial and temporal order emerges spontaneously, in the absence of a central directing agency, in a complex molecular system kept away from equilibrium by a flux of energy. Self-organization is not a mechanism; it is the label for a pigeonhole that will hold the relevant forces and rules of engagement. In the case of Bènard's pan of oil, the key is the replacement of diffusion by convection as the mechanism of heat transfer. The behavior of a cellular system is harder to comprehend, but growth and morphogenesis in *E. coli* ought to be a far more tractable case than the development of an embryo.

The beginning of practical wisdom in this matter is, I believe, to be found in quite traditional concepts. Every cell comes from a parent cell, which provides a templet upon which the daughter cell is modeled; and the parent cell's epigenetic landscape acts in concert with its genes to guide the process of reproduction. New gene products (and, for that matter, products of biosynthesis) are generated in a context that already possesses a degree of spatial structure, and they take their places in an existing order. It is true, as Shapiro and Losick (19) put it, that "cell fate is established by proteins that localize to the site of cell division." But that localization process, whatever its mechanism proves to be, is itself a reflection of global cellular organization. It displays (but does not ultimately cause) the orderly, channeled flow of metabolites, gene products, energy carriers and signals through space and time. Spatial markers, such as the pole/wall junction of *E. coli* and the conspicuous wall bands of streptococci, are some of the effectors and indicators of that spatial order. There will surely be others: one can imagine localized signaling complexes, and the progressive hierarchical elaboration of spatial order upon that initial fixed point. Soap-bubble physics, materializing a shape that is compatible with some particular set of genetically specified boundary conditions, illustrates the next higher level of collective order: *if* turgor pressure and a parameter corresponding to surface tension fall in the right range, *and* a rigid support is in place at each end, *then* a cylindrical shape will emerge. Other specifications enter into the morphogenetic calculus, some known and others unknown. And so here comes the challenge to the next generation of researchers: now that we know most of what is worth knowing about the genetic level of biological organization, it is the epigenetic spaces that hold the best prospects for significant new discoveries.

Consider, for example, the matter of morphogens. The discovery, that the developing embryo of *Drosophila* (and of other animals) is blocked out by a grid of gradients formed by diffusible proteins called morphogens, is surely one of the shining achievements of latter-day biology. Are diffusible morphogens a clever trick of the animal kingdom, or do they represent a new universal of biology? Thus far, what we have learned of bacterial development and morphogenesis suggests that local, short-range interactions suffice to shape the cell, but do they? Would it be useful to think of a bacterial cell as a morphogenetic field, whose numerical parameters are set by gene products? Is the shape of a bacterial rod, evidently the systemic property of a collective of genes and proteins in a spatially organized context, predictable in principle as one solution of a set of differential equations that prescribe its self-organizing behavior? I don't know the answer to any of these questions. What I do know is that, were I again a young investigator, I would not be content to bash promoter sequences; I'd want to go and look behind the ranges, where something new may be hidden.

7

MORPHOGENESIS: WHERE FORM AND FUNCTION MEET

"The main battle ground . . . is the problem of the relation of function to form. Is function the mechanical result of form, or is form merely the manifestation of function or activity? What is the essence of life, organization or activity?"

E. S. Russell (1)

"There is a clear research programme for the study of biological morphologies as natural forms, as attractors in the space of morphogenetic field dynamics."

B. Goodwin (2)

DYNAMIC ARCHITECTURE OF EUKARYOTIC CELLS
FORCE AND COMPLIANCE
APICAL GROWTH
FIX ME AN AXIS
AMOEBOID CELLS: FORMS IN FLUX
ARRANGING A CILIATE
IN PURSUIT OF WHOLENESS

Biology began as an observational science, with its focus on the morphology of organisms and their parts; as far as the general public is concerned, the fascination of the living world still springs from the variety and beauty of its forms. And why not? We are visual creatures and know our world first and foremost by what we can see; were it not for those forms we could not recognize organisms or reflect upon their nature.

Scientists tend to view the subject from another angle. In the nineteenth century, following the rise of evolution as a unifying framework for all of biology, function displaced form as the wellspring of inspiration; a trend reinforced in the twentieth century, as biology metamorphosed into an experimental science. Physiology, biochemistry and now molecular biology revolve around the activities of organs, cells or molecules, described very largely in chemical terms. The shapes of these entities are apt to be brushed off as secondary features from which nothing fundamental can be learned.

Like E. S. Russell in his day, I would not hide my "sympathy with the functional attitude," and particularly with the belief that variation and selection for functional performance (rather than overarching laws of form) are the chief architects of the living world. But the dismissal of form from our universe of discourse has gone too far, for in truth form and function, physical law and historical constraint are intertwined at all levels from the molecular to the organismic and the communal. I am therefore encouraged by the recent revival of interest in the relationship of function to form, and particularly in the processes that underlie biological morphogenesis. Quietly and without media fanfare, a minority of biologists are looking beyond the genetic paradigm and rediscovering organisms. To us, as to our predecessors before Darwin, organisms are "wholes": self-maintained and persistent patterns in space and time. From this viewpoint form becomes central once again, the primary expression of an organism's identity and a point of departure for the exploration of the higher levels of biological order.

Reflect for a moment on the sampler of eukaryotic unicellular organisms depicted in Fig. 7.1, which will appear in various contexts in this and the following chapters. How do these shapes arise as each cell grows, divides, and traverses its life cycle? How are they transformed in the course of development, or regenerated after injury? How are these forms so faithfully transmitted, generation upon generation, that a glance is often sufficient to identify a species and one can recognize *Paramecium* in amber 60 million years old? Can one discern the forces that hold some forms constant for aeons while others are transfigured in a geological instant? And what do these forms mean: are they products of natural selection, frozen accidents of biological history, or expressions of morphogenetic laws? Such riddles sketch the scope of contemporary research into the nature and significance of cell morphology. The goal, which is just coming into focus, is to go beyond the description of events supplied by generations of microscopists, and now being amplified by dissection to the molecular level. We strive to make these forms intelligible as manifestations of coherent, causal processes on the cellular

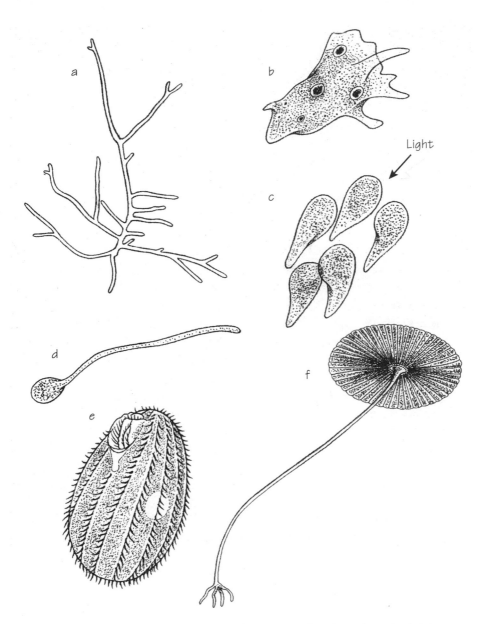

Light

Fig. 7.1. A gallery of eukaryotic microorganisms. (a) Young fungal mycelium; hyphal diameter 10 μm (b) An amoeba, *Mayorella*, on the prowl; length 25 μm (c) Embryos of *Pelvetia* developing synchronously under unilateral light from the top of the page; length 300 μm (d) Germling of an oömycete; hyphal diameter 10 μm (e) The ciliate *Tetrahymena*, with mouth and rows of cilia; length 30 μm (f) The giant unicellular green alga *Acetabularia mediterranea* with holdfast, stem and reproductive cap; length several centimeters.

scale (micrometers to millimeters), governed both by local molecular interactions and by global organizing principles.

Morphology and morphogenesis enter into the physical sciences (think of crystals, or the way glaciers sculpt mountains), but biology is their particular province, thanks to the intimate connection between form and function. Living shapes, like non-living ones, are produced by physical forces and conform to the laws of physics, but in the living, purpose informs and constrains physics. The fact that DNA is a long linear string while proteins are roughly globular with rugged surfaces is simultaneously cause and consequence of their biological functions, the molecular morphologist must be concerned with history no less than with chemistry. The same should be true at the organismic level; but the complexity of the patterns obscures the connection between chemistry and history. The generation, persistence, and evolution of large-scale spatial patterns defines the central issues in biological development, a subject that remains today almost as baffling as it is fascinating. In keeping with the bounds I have set for this book, this chapter treats of cells and unicellular organisms rather than the more "relevant" flies and frogs. For nearly two decades my laboratory was engaged in research on form and function in eukaryotic microorganisms; some overlap with earlier articles (3) is unavoidable, but I hope that time has sharpened my perceptions.

Research on morphogenesis, like most other branches of biological inquiry, has lately turned obsessively molecular. The point of faith, seldom stated yet plain enough, seems to be that when our understanding of the molecular parts is complete, organizing principles will stand forth more or less automatically. The reader will already have observed that I do not share this faith. On the contrary, it seems to me self-evident that a causal, dynamic and comprehensible account of cellular order calls for the marriage of the bottom-up view of molecular chemistry with the top-down perspective of physiology, the science of complex systems. The object of this chapter is to consider how far the courtship has advanced. With the focus on the articulation of the pieces rather than their particulars—on forces, flows and vectors—it stands squarely in the tradition of cell physiology. A single thread runs from D'Arcy Thompson through Conrad Waddington and Tracy Sonneborn, to contemporaries such as Joseph Frankel, Brian Goodwin, Lionel Harrison and Lionel Jaffe. I am proud to acknowledge my debt to these and other integrative scholars, for their view of life is also my own.

DYNAMIC ARCHITECTURE OF EUKARYOTIC CELLS

Eukaryotic cells are to prokaryotic ones what a mansion is to a studio apartment. The studio is functional for all purposes but optimal for none. The mansion, larger and more elaborate, has chambers set aside for kitchen, library and play room. Just so, eukaryotic cells contain a true nucleus with chromosomes and a nuclear membrane studded with pores, organelles for energy generation, an extensive network of endomembranes, and a pervasive cytoskeleton. What is all this construction good for? Part of the answer is simply that eukaryotic cells are usually much larger than prokaryotic ones. Bacterial cells are small enough to put all their genes on a single string and to let diffusion handle the distribution of metabolites and gene products. Eukaryotic cells (with a few exceptions) have ten times as many genes and protein species per cell, a thousand-fold larger volume but a surface-to-volume ratio only one twentieth that of bacteria. Their design requires the plasma membrane to specialize in communication with the outside world, while internal organelles take over the task of energy generation; and a scaffold of girders, cables, tracks and tubes ties all cellular operations into a coherent unit.

I do not mean to imply that eukaryotic cells are the product of intelligent, purposeful design, the supposition is that the adaptive evolution of a cytoskeleton and intracellular membranes made possible the proliferation of larger cells displaying varied and elaborate morphology. All the same, it is instructive to examine eukaryotic cells from the viewpoint of design ("reverse engineering"), as diverse ensembles of parts that answer to particular constraints and serve functional purposes. The function with which we are here concerned is the construction of complex forms on a scale far above the molecular, from micrometers to millimeters (and beyond: the neurons that control the giraffe's neck must be several meters in length). In the generation of large-scale order, internal membranes and the cytoskeleton play the star roles.

Eukaryotic cells harbor two classes of internal membrane-bound structures. The first comprises the discrete organelles: mitochondria, chloroplasts, peroxisomes and (in certain anaerobic protists) hydrogenosomes. These are structurally and functionally related to prokaryotes, and are now known to be descendants of ancient bacterial endosymbionts (Chapter 8). The subject here is the second class, the dynamic endomembrane system, which includes the nuclear membrane, endoplasmic reticulum, Golgi, lysosomes and vacuoles. These recognizable and more-or-less permanent elements are connected by a ramifying network of tubules and vesicles that fills much of the cytoplasmic volume and links the central elements to one another and to the plasma membrane. A three-dimensional cartoon (Fig. 7.2) depicts the eukaryotic cell as a diaphanous maze of membranous surfaces that define compartments of specialized function and composition, all of which lie topologically outside the cytoplasm proper and connect, at least intermittently, with the exterior. This still does less than justice to the dynamic nature of the endomembrane system. What we refer to as ultrastructure is more like a snapshot of a fluctuating flow of membranes and substances associated with them, from the nucleus to the periphery and from the surface into the center.

The functions of the endomembrane system revolve chiefly around the traffic in macromolecules and larger objects. Cells that live by ingesting prey, as ciliates do, must internalize and digest food particles; the endomembrane system is part of the apparatus by which cells eat (and, for that matter, drink). Genes are housed in the nucleus, but their products must often be despatched to distant locations; the endomembrane system effects their delivery. One of its primary functions is the secretion of proteins and polysaccharides, including the precursors of cell walls. In walled cells, localized secretion is commonly the key to growth and morphogenesis, and the spatial organization of the endomembrane system itself is integrated with the physiology of the whole organism.

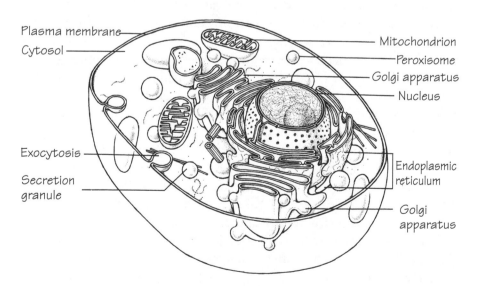

Fig. 7.2. A maze of membranes. After de Duve, 1984, with permission of the de Duve Trust and W. H. Freeman and Company.

The advantages of transporting goods by means of sealed and addressed containers must have been discovered early in the evolution of the eukaryotic cell, since it is one of their hallmarks. By way of illustration, consider the path of some protein destined for secretion into the medium (Fig. 7.2). The cognate gene is transcribed in the nucleus, and the messenger RNA is exported into the surrounding zone of endoplasmic reticulum. Ribosomes attached to those membranes translate the message, and concurrently extrude the growing protein chain into the lumen of the maze. Having traveled along those tubules, the protein is incorporated into a particular class of vesicles that bud from the endoplasmic reticulum and carry their cargo to the next way station, the Golgi complex. The Golgi, commonly a conspicuous and ornate stack of membrane-bound vessels, serves as the distribution hub of the secretory pathway. There, proteins are sorted and undergo chemical modifications that address them to their ultimate destination. Successive vessels in the stack may each carry out a particular stage, and they are commonly linked via mobile vesicles that bud from one vessel to fuse with the next. Eventually the protein, duly addressed and processed, is carried by yet another class of vesicles to the plasma membrane, where secretion proper takes place. Exocytosis entails the fusion of secretory vesicles with the plasma membrane such that the cargo protein is released to the outside. Exocytosis is commonly regulated by the cell, which controls both place and time.

There is presently keen interest in the molecular mechanisms that mediate all facets of this process: protein translocation across membranes, the architecture and formation of transport vesicles, the mechanisms of fusion and their regulation. Details will be found in any textbook of cell biology and in a steady stream of specialized review articles. Matters of form and overall organization of the system are not nearly so popular. How, for instance, do these famously fluid membranes hold their shape? How does the endomembrane system as a whole acquire its spatial orientation and location, while the cell of which it is a part grows, divides and moves around? The short answer is that the endomembrane system is strung upon the cytoskeleton like laundry on a line, and thereby partakes of the dynamic, long range-order of the cell as a whole.

In a nutshell, the cytoskeleton is responsible for the mechanical integration of cellular space; unpacked, this phrase covers a host of actions and interactions, mediated by a large and growing ensemble of proteins. Some of the cytoskeleton's functions would be performed on a building site by scaffolding, girders and struts, tracks, cables and pulleys. It supplies mechanical support as well as motive power: "cytobones" and "cytomuscles," as Christian de Duve says. The traffic in vesicles outlined above is spatially organized by a spiderweb of rails, along which vesicles travel with the aid of motor proteins powered by ATP. Organelles, including the endomembrane system, keep their place because they are pegged out on the tracery of cytoskeletal filaments.

In cells lacking a wall, such as animal cells in tissue culture, the cytoskeleton also serves as the structural framework. Passively rigid elements balance active contractile force exerted by the periphery, and it is the opposition of force and resistance that hold up the cell. There seems to be no satisfactory analogy for this sort of construction from everyday life, but shrinkwrap suggests the principle. A grander parallel, drawn by Ingber (4), is to tensegrity-based architecture, which is supported by the interplay of tensile and rigid elements.

Structural mechanics is one aspect, dynamics the other, for the cytoskeleton holds together thanks to the continuing expenditure of energy, and it is subject to frequent remodelling. Mitosis, for instance, entails the dissolution of much of the cytoskeleton; its components are redeployed in the service of cell division, and subsequently reconstituted in their former order. Everything is in flux, but in a regulated purposeful manner. Neither morphogenesis nor any other cellular activity of eukaryotic cells can be understood apart from the cytoskeleton.

What girders and cables are to the construction business, microtubules and microfilaments are to the cytoskeleton. Complementary in mechanical properties and functions, both are required to delineate and integrate cellular space; they are universal architectural elements of eukaryotic cells, but absent from prokaryotes. As the name implies, microtubules are hollow tubes made of parallel strands of polymerized tubulin (usually 13 strands, for a diameter of 25 nanometers). They are strong, sometimes rigid, and resist both shearing and compression as steel tubing would. Microfilaments, double strands of polymeric actin, are thinner (diameter 6–7 nanometers) and flexible; they exert or resist tension, as string does. In the cell, single microfilaments are commonly twisted together into thicker ropes. Microtubules make scaffolds, microfilaments go into nets, meshwork and cables, and both can provide tracks for the transport of cargo.

Actin is one of the most abundant of cellular proteins and very nearly universal. Tubulin is less abundant but familiar to every biology student as the substance that mitotic spindles are made of. Both are common and versatile structural materials, shaped into sundry patterns and put to manifold uses with the aid of a galaxy of "ancillary" proteins. It is the latter that should receive credit for the specialized tasks performed by the cytoskeleton on behalf of its parent cell. Ancillary proteins stabilize filaments, sever them and link them to each other and to structures beyond. Motor proteins have become a growth industry: we know more than 30 varieties of kinesin, a microtubule-associated motor protein, and a comparable number of myosins, which power movements of and along actin filaments. These and other processes are regulated by an ever-growing number of signals, each with its own set of protein receivers, transmitters and switches. The molecular basis of cytoskeleton functions is presently one of the most intense areas of research in cell biology, but it falls outside the scope of a chapter concerned with the higher levels of order.

The molecular elements of the cytoskeleton, nanometer-sized grains, find their places in cellular structures and functions larger by three to five orders of magnitude. This supramolecular level of organization has its own characteristic building blocks and patterns. Take microtubules. The tubule proper, a self-assembling complex of $\alpha\beta$ tubulin dimers, defines the first level, but the time and place of self-assembly are anything but random. Growth of a microtubule (more likely, a bundle of them) is nucleated upon a specialized structure, such as the basal bodies from which cilia sprout or the centrosome of the mammalian cell. Such entities are designated microtubule-organizing centers (MTOCs in the trade), and the tubule grows out in a polarized manner such that assembly occurs chiefly at the distal end. So does disassembly, and a growing microtubule is apt to shrink periodically as it explores its surroundings. But when it encounters a capture site (on a chromosome during mitosis, or at the cell membrane), the microtubule is stabilized and becomes fit for mechanical duties. Microfilaments assume the form of regular or irregular meshwork, thick or thin fibers, solid plaques or portable fragments, all in response to changes in pH, calcium concentration or other signalling molecules. One cannot predict the form or function of these complex ensembles from the characteristics of their component proteins, but when seen in the context of the parent cell the arrangement of the molecules becomes quite comprehensible.

From the viewpoint of the biochemist or cell physiologist, how the cytoskeleton works and grows shapes the cell. But how is the cytoskeleton itself so fashioned that its operations accord with the cell's overall "plan" and generate its particular morphology time after time? Cellular organization is the product of a hierarchy of operations. When we understand how global order emerges from the succession of local interactions, morphogenesis will still present challenging puzzles; but it will be a mystery no longer.

Force and Compliance

On Growth and Form by D'Arcy Thompson was first published in 1917, and has never since gone out of print. This is a rare distinction, shared with Darwin's *Origin of Species* but with few other books of science. The juxtaposition is not altogether capricious, for Darwin and Thompson represent opposite and complementary ways of looking at the living world. Thompson did not reject the idea of evolution by natural selection; it just did not much interest him, for it offered little insight into the generation and meaning of biological forms. Physics, on the other hand, clearly did. "The form . . . of any portion of matter, whether it be living or dead, and the changes of form which are apparent in its movements and in its growth, may in all cases alike be described as due to the action of force. In short, the form of an object is a 'diagram of forces' in this sense, at least, that from it we can judge or deduce the forces that are acting or have acted upon it" (5). We can still read Thompson with pleasure and profit to learn about the shapes of seashells, leaves and horns; the angles drawn by the spicules of sponges or the partitions in the bee's honeycomb; and how by simple rules the carapace of one crab can be transformed into that of another. We must also adopt Thompson's viewpoint if we hope to understand how the shapes of molecules, organelles and cells come about, for they too are produced by physical forces (5). The challenge is to identify the forces in question, and to determine how their action is localized in space and time.

Thompson devoted a single chapter to the forms of cells and unicellular organisms, which turns on the proposition that many biological shapes result from surface tension. This force is a consequence of the mutual attraction between molecules in a liquid, and acts parallel to the surface. In objects governed by surface tension, the shape will be that which has the smallest surface area compatible with volume and with mechanical constraints. The sphere is the simplest and most familiar shape of this kind, but by no means the only one. We noted in the preceding chapter that soap bubbles, whose properties are determined by surface tension in the film of soapy water, can be made to assume diverse shapes with the aid of structural supports: a bubble blown upon a pair of rings can be cylindrical as long as its length does not exceed its circumference. Spheres and cylinders are common among microorganisms, and Thompson noted instances of more esoteric forms built upon parabolas and ellipses that obey the rule of minimal surface. Is it surface tension that shapes the spherical egg or the flattened fibroblast?

Not surface tension *sensu stricto*, but something like it. In his editorial introduction to Thompson's chapter, John T. Bonner points out that it was already clear in Thompson's day that the surface tension of lipid membranes is too small to determine cell shape. However, cell surfaces are commonly subject to contractile forces operating at the level of the underlying meshwork of actin filaments. From the viewpoint of cellular morphology, the physical nature of the force is less important than the general observation that surface layers often behave as though they were fluid, and assume a configuration that minimizes the surface area for any given volume and mechanical constraint. The physicist's surface tension is a measure of the amount of work that must be done to enlarge the surface by an incremental unit, and it is a constant for any particular surface. The biologist's "envelope tension," is likewise a measure of the work required to overcome the sum of cohesive forces, but may vary as a function of location and of the metabolic status at any given time. We encountered one application of this principle in the preceding chapter, and it will recur in other guises below.

Cells are also subject to a global countervailing force, namely hydrostatic pressure that tends to enlarge the volume and expand the surface. This arises because, as a rule, the concentration of molecules and ions in the cytoplasm exceeds that of the medium; water therefore tends to flow inward, expanding the cytoplasmic volume. What happens next depends on whether or not the cell is surrounded by a strong wall. In most bacteria, fungi, algae and plants the wall resists enlargement of the volume. The resulting hydrostatic pressure, or turgor pressure, stresses the wall and is generally thought to be part of the mechanism by which walled cells enlarge; this was discussed in the bacterial context in the preceding chapter. For cells not protected by a wall, such as those of animals and many protists, the influx of water threatens swelling and eventually rupture. They circumvent the problem in one of several ways. Cells living in relatively concentrated media, such as blood serum or sea water, maintain fluid balance by continuously extruding salts into the medium; this minimizes water influx by reducing the osmotic gradient between the cell and its surroundings. Cells that live in fresh water, and many others cells as well, regulate the concentration of cytoplasmic solutes; some even expel fluid from the cytoplasm with the aid of a device not unlike a kidney, called a contractile vacuole. All the same, a small hydrostatic pressure still persists even in unwalled cells, balanced by envelope tension; whether this pressure plays a part in cell enlargement and growth is not clear.

Finally, cellular structures that impinge upon the cell surface may exert direct force upon it, or reach across the surface to the extracellular matrix. Animal cells commonly maintain their normal morphology only as long as they are attached to a solid surface and continuously pull upon it; cells that have lost their moorings tend to become spherical. Attachment is mediated by specialized adhesion pads that connect across the plasma membrane to cellular filaments, where the mechanical force is produced. Walled cells are not so different from unwalled ones: here, too, there is evidence that cytoskeletal fibers reach across the plasma membrane, to contact the wall and exert tension upon it. This illustrates the dual roles of the cytoskeleton in cellular morphogenesis: it generates mechanical force, and it applies it in a highly localized manner.

One of my prized possessions is an exquisite flint knife, knapped five thousand years ago on the coast of modern Israel; its maker knew just how to set his tool so that localized force would detach exactly the right spall. A glassblower, by contrast, uses his torch to soften a particular spot and lets pressure blow the bubble; his kind of morphogenesis depends on localized compliance. Cells draw on both skills to shape themselves, and it is usually the cytoskeleton that localizes the effect. Beyond this broad abstraction, variety is the rule, for the question, "How do cells shape themselves?" has no unitary answer (as questions about molecular mechanisms often do; Chapter 4). Students of morphogenesis seek causal and dynamic accounts of the events that shape particular cells or organisms; and these reports, each of them unique, will be narratives about forces, compliance and modes of localization. Let us look at some examples.

APICAL GROWTH

The soil by the old stump is laced with thin, white threads: fungal hyphae, silently reducing the rotting wood to nutrients, energy and more fungus. We do not much notice fungi until their fruiting bodies pop up in the lawn, but were it not for their diligence, all organic carbon would long since have been tied up in dead plant matter. The capacity of fungi to explore solid substrate, invade tissues and secrete digestive enzymes are all linked to their particular mode of growth, extension at an apex or tip. Some organisms unrelated to fungi do likewise—actinomycetes, algal filaments, pollen tubes, root hairs and neurons come to mind—but in the case of the fungi, apical growth holds the keys to the kingdom.

The hypha is the archetype of fungal form, a tube of approximately constant diameter ending in a rounded or tapered tip; the filament may or may not be subdivided by septa into individual cells (Fig. 7.1a). Hyphae also provide the paradigm of apical growth: extension is confined to the tip, which makes up a minute fraction of the length. When conditions are favorable, hyphae branch, putting forth new tips from regions that were previously quiescent. The branches extend and proliferate, in turn, giving rise to the familiar pattern of a fungal colony spreading over a slice of bread. The adaptive value of this mode of growth is self-evident: like motility, apical growth continually makes fresh nutrients available to the growing zone at the hyphal front. Fungi take on other forms as well, but most can be regarded as modified filaments; even the familiar mushroom develops as an aggregate of innumerable tip-growing hyphae. As in the case of bacteria, the form of the hyphal tube is maintained by a strong and rigid wall, but does not obviously depend on the chemical composition of that wall: morphologically, the chitin-based hyphae of the true fungi differ only in detail from the cellulose-rich hyphae of oömycetes or pollen tubes. Hyphal shapes can also not be credited to an orderly sequence of subunit assembly, for hyphal walls consist of an irregular meshwork of fibrils (chitin or cellulose, as the case may be) embedded in an amorphous matrix in which little order can be discerned. What fungal hyphae have in common, and share with other tip-growing organisms, is the manner in which wall deposition is organized in space. The generation of metabolic energy takes place all along the hypha's length; so does the synthesis of macromolecules and the production of organelles, all of which are brought forward by cytoplasmic flow. But new cell wall and plasma membrane are laid down almost exclusively at the far end. Hyphal morphogenesis, the production of a growing tube with tapered tip, revolves around what takes place in that narrow corner.

The ground-breaking insights into the physiology of hyphal extension came from ultrastructural studies carried out three decades ago by M. Girbardt in Germany, C. E. Bracker in the United States and P. M. Robinson in the United Kingdom. What they found is that a hypha is a highly polarized secretory cell (Fig. 7.3). The apex is devoid of mitochondria, nuclei and the other standard organelles, but is filled with small membrane-bound vesicles. In some organisms, notably the "higher" fungi (*Neurospora* and *Sclerotium* are examples), the apical vesicles are collected into a coherent structure called a Spitzenkörper, or apical body (Fig. 7.3a). In others, including the oömycetes, yeast and also pollen tubes, vesicles cluster at the tip more or less at random (Fig.

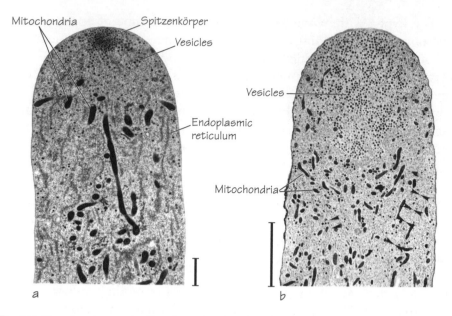

Fig. 7.3. Cytoplasmic organization of growing tips. (a) A hypha of the fungus *Sclerotium rolfsii*. Note the apical zone devoid of organelles but filled with secretory vesicles; the Spitzenkörper(s) is a dense aggregate of vesicles. Bar 1 μm. From Roberson and Fuller, 1988, with permission of Springer Verlag, Wien, Original photograph courtesy of Dr. Robert Roberson. (b) Tip of a pollen tube, filled with secretory vesicles but without a Spitzenkörper. Bar, 5 μm. From Lancelle et al., 1997, with permission of Springer Verlag, Wien.

7.3b). As the hypha grows, the vesicles fuse with the plasma membrane at the extreme tip (the outermost two micrometers) and their contents are secreted into the wall space beyond. Apical vesicles come in several varieties; their contents are not fully established, but we can take it that the vesicles deliver membrane lipids, precursors and enzymes for wall synthesis, and also digestive enzymes. It is the subsequent, highly localized production of new wall and membrane that shapes tip and tube. Many of the mechanistic details remain to be worked out; the condensed account that follows can be amplified by reviews of the technical literature (6).

One open question that bears directly on the directional character of fungal growth is just how the secretory vesicles, generated along the hyphal trunk in organelles equivalent to the Golgi apparatus, are brought forward to the tip. There is evidence that both microtubules and microfilaments serve as tracks for vesicle transport, but these do not usually extend into the apical dome itself. The Spitzenkörper, when present, serves as a way station for vesicles in transit to the apical membrane, but we do not know what transpires in that body, nor what takes its place in those organisms that lack the Spitzenkörper. What is established is that vesicles undergo exocytosis at the extreme apex, and that new wall synthesis is restricted to that location. Nascent tip wall is plastic, but hardens on a timescale of seconds as the fresh wall increment falls behind the advancing tip, undergoes stretching, and becomes part of the hyphal trunk.

What makes the tip advance? If it did not, apical secretion would merely thicken the wall. The conventional answer is that turgor pressure supplies the driving force, continually distending and shaping the nascent plastic apex. Hyphal extension thus illustrates the general idea that morphogenesis results from local compliance with a global force, in this instance hydrostatic pressure. One would then expect turgor pressure to be required for hyphal extension, and that appears to be true for higher fungi. Surprisingly, however, the hyphae of certain oömycetes extend and shape a tube with rounded tip even when their turgor pressure has been artificially lowered to the point where it is no longer measurable (7). Extension is physically possible because hyphae growing with diminished turgor deposit an increasingly plastic wall. It follows that vesicle transport, extension and the spatial pattern of exocytosis are fundamentally independent of turgor pressure; in oömycetes, at least, the advance of the tip may be driven by ATP-linked motor proteins or by extension of the cytoskeleton. The crucial question then becomes how vesicle fusion is tightly localized to the apex, for this ensures the localized expansion of the freshly-deposited wall.

The short answer is that we do not fully understand how exocytosis is restricted to the apex. A longer one begins with the recognition that tip-growing organisms may not all employ the same mechanism, and that the discussion revolves around three distinct possibilities (8). First, vesicles may be delivered right to the locus of exocytosis by dedicated tracks, like packages to a railway station. Sites of wall expansion are commonly also sites of microfilament assembly (during budding of yeast, for instance), and one function of these microfilaments could be vesicle transport. Recall, however, that in both fungal hyphae and pollen tubes cytoskeletal fibers are largely excluded from the apex proper, so this cannot be a universal mechanism. A second idea, presently very prominent, calls for the locus of exocytosis to be identified by a cellular signal; the favorite is a localized influx of calcium ions, mediated by short-lived membrane channels carried in the secretory vesicles themselves. This hypothesis, whose development was begun by Lionel Jaffe thirty years ago, has now received powerful support from research on the extension of plant pollen tubes (8). In this case it is well established that calcium ions flow from the external medium into the apex, and that the calcium concentration in the apical cytoplasm is ten-fold higher than elsewhere. Growth of the pollen tube is tightly correlated with this gradient of cytosolic calcium; for example, artificial displacement of the calcium peak displaces the direction of growth. Just what the calcium ions control is uncertain, they may promote vesicle fusion at the apex, but may have other tasks as well. Evidence from oömycete hyphae points to a similar mechanism of localization, but in the case of true fungi there is (presently) no compelling evidence for localization of vesicle exocytosis by calcium ions.

For these latter organisms, S. Bartnicki-Garcia and his colleagues have proposed a radically different mechanism of localization that turns on the Spitzenkörper. In their view (9) the Spitzenkörper serves as the immediate source of the vesicles that fuse with the plasma membrane, and is itself mobile (possibly with help from the cytoskeleton). The Spitzenkörper operates like a moving nozzle that spray-paints the wall ahead. With the help of some simplifying assumptions, these authors derived an equation that predicts the shape a hyphal tip should display, given only the rate and direction of Spitzenkörper movement and the rate of vesicle production. In many cases the calculated form corresponds uncannily well to the observed one, and several recent experiments in which the position of the Spitzenkörper was perturbed or manipulated strongly support the idea that this organelle is the primary determinant of tip shape. Whether or not the details hold up to the scrutiny that is now under way, the development of a quantitative and physiologically explicit hypothesis that correctly predicts a biological shape is surely a remarkable achievement.

Much remains to be learned, but the basic principles of apical growth and morphogenesis seem to be taking shape. The keys are one or more forces to drive enlargement, and focused secretion to support localized expansion of the cell wall. The nature of the driving force, just how the graded distribution of fusion sites comes about, and the chemical and physical parameters of the nascent wall can all vary from one organism to another. But diversity is subsumed under an overarching unity. So long as the sites of exocytosis are arrayed in a steep gradient and advance vectorially, and so long as nascent wall is plastic but stiffens with time, apical growth must ensue. For apical growth, at least, a general explanation of morphogenesis is in sight.

Fix Me an Axis

"Growth and morphogenesis represent transport processes, which in common with the more popular membrane transport, must be described by vector quantities, having both magnitude and direction." Peter Mitchell (10) made this point at an early stage in the development of his ideas on the spatial direction of biochemical reactions (Chapter 5), but never pursued it further. It remains an unconventional viewpoint but an instructive one, for almost everything that happens in a living cell or organism has a direction in space. At the cellular level we speak of "polarity," the vectorial quality so forcefully illustrated by the physiology, behavior and growth of fungal hyphae. But where does the ubiquitous polarity of cells come from? As a general rule, a cell inherits its polarity from its progenitor(s) in the course of cell reproduction. But there are instances in which polarity arises *de novo* from an unpolarized state, and these afford insight into the nature of physiological vectors.

The marine brown algae *Fucus* and *Pelvetia* are familiar to anyone who tramps the rocky shoreline between California and Alaska, but only biologists are likely to know that the germination of their fertilized eggs has been a system of choice for the study of cell polarity for more than a century. Briefly, the plants discharge eggs and sperm into the sea, where fertilization takes place. The resulting zygote appears spherically symmetrical, both structurally and physiologically, and it remains symmetrical as the zygote lays down a primary cell wall, attaches to a rock, accumulates salts and begins to generate turgor pressure. But then polarity asserts itself (Fig. 7.4). The embryo develops a bulge that turns into an outgrowth, which defines the polar axis of the adult plant. The first cell division, at about 22 hours, demarcates two unequal cells that have different developmental fates. The rhizoid cell, comprising the outgrowth, is destined to become the holdfast; in the early stages of development this is the growing part, for the embryo's first object is to secure firm lodging. The larger thallus cell is destined to become the frond. What makes this process so attractive for study is that the symmetrical zygote develops polarity by reading directional cues from the environment. In nature the chief cue is unilateral light: embryos germinate on the shaded side (where the rock is; Fig. 7.1c) and this initial polarizing event is progressively amplified and fixed as development proceeds.

The ultrastructure of the germinated embryo gives one a good sense of what is meant by a polarized cell (Fig. 7.4b) and its axis of

a. Germination   b. A polarized embryo

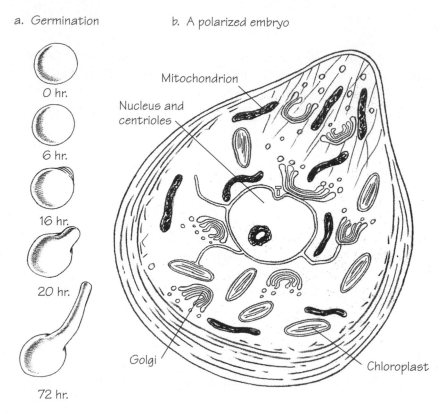

O hr.

6 hr.

16 hr.

20 hr.

72 hr.

Mitochondrion

Nucleus and
centrioles

Golgi

Chloroplast

Fig. 7.4. Germination and outgrowth of the zygote of brown algae. (a) Gross morphology at the time of fertilization and at 6, 16, 20 and 72 hours afterward. The first signs of asymmetry appear at about 15 hours. Diameter 150 μm. (b) Ultrastructure of a polarized germling showing oriented nucleus with centrioles, mitochondria and Golgi facing the site of outgrowth; and cytoskeletal tracks guiding vesicles to the apex. This is a highly speculative drawing, based with permission on an unpublished sketch by Susan Brawley.

development. In the egg, even after fertilization, organelles and cyto-skeleton are symmetrically arranged. As the embryo germinates, assymetries develop progressively over a period of hours, reaching into ever higher levels of cellular order. The sequence of events and the causal relationships that link one to another have been under intense investigation since the sixties, and the outline is becoming clear (11). The first signs of structural polarity appear after three to four hours of exposure to unilateral light, beginning with the assembly of a ring of microfila-ments in the cortex of the zygote's shaded half; the ring marks the site where secretion begins, and where outgrowth will eventually take place. About the same time calcium-channel proteins accumulate in that re-gion, and a current of calcium ions begins to flow into the cell. Calcium ions accumulate locally, perhaps promoting vesicle fusion, and additional microfilaments migrate to the site. A cortical complex of cytoskeletal and channel proteins, whose formation precedes visible outgrowth and predicts its location, marks the target site for the directed secretion of Golgi vesicles. The early stages of cell polarization ("axis formation") can be reversed by changing the direction of incident light; but as secretion proceeds the embryo's axis becomes permanently fixed.

Microfilaments play a prominent role at all stages of polarization, which can be blocked by drugs that inhibit microfilament assembly. Actin is a structural component of that cortical complex which determines and stabilizes the locus of secretion. It is thought to contain, besides microfilaments, elements of the cell wall, proteins that connect the wall with the cytoskeleton, and components of a signal-transduction cascade that includes a local peak of calcium concentration. Do microfilaments constitute the tracks that guide secretory vesicles to the site of exocytosis? Is the cytoskeleton physically linked to the cell wall at that site? These are just two of the questions that seem sure to be settled by the application of sophisticated methods now being developed to light up particular proteins in the living organism and to manipulate their functions.

Once localized secretion is under way, outgrowth begins; we can envisage it as a kind of apical growth, powered by turgor pressure and localized by targeted exocytosis of secretory vesicles that deliver precursors and enzymes for wall expansion. Concomitantly, polarization spreads to higher levels of order. Mitochondria become aligned upon the axis, microtubules sprout from the centrosomes and reach into the tip, the nucleus moves forward and rotates; eventually the nucleus divides and is bisected by a cell plate, transverse to the axis, to separate the rhizoid cell from the thallus. Polarization is a process, not an event; its heart is the establishment of a permanent vectorial secretory pathway that links the nucleus to a particular region of the periphery.

In the foregoing description, unilateral light supplies the vector that determines the direction of outgrowth and fixes the developmental axis. This is surely true in nature, but there is no such obligation in the laboratory. Fertilized eggs germinate quite normally in the dark, responding to any one of a variety of gradients: they germinate towards the high end of a K^+ gradient, the acid end of pH gradient, the high end of a gradient of calcium ionophore. Even the absence of any external gradient does not stump them: the site of sperm entry becomes the locus of outgrowth (indeed, the first tiny ring of microfilaments assembles at that site, and is subsequently repositioned in accordance with environmental cues). These observations are not mere laboratory curiosities, but tell us something significant about the origin of spatial order. "There is something about the internal dynamic organization of an egg that makes spherical symmetry unstable, so that any perturbation . . . will get it started. What light and other stimuli do is influence *where* the axis will appear, but it is not the cause of axis formation itself. It is just a trigger that initiates something that is poised and ready to go, like a sprinter at the start of a race." (12).

Every organism shapes itself in its own way, but as our knowledge grows, commonalities emerge that cut right across the living spectrum. Not only do organizational patterns recur, but the molecules involved are often homologous (related by descent from a common ancestor). Bud formation in the yeast *Saccharomyces* (a fungus), just like germination of algal zygotes, calls for the localized expansion of the cell wall at a selected site. The chemistry of yeast wall is quite unlike that of *Pelvetia*, and the bud site is usually marked by a tag left over from the previous division cycle rather than an environmental vector. But what happens next bears a clear resemblance to the germination of *Pelvetia* as outlined above (13). It turns on the assembly of a cortical complex containing microfilaments and the components of a signalling cascade; this becomes the target patch for the directed secretion of vesicles, and also regulates the activity of enzymes that deposit cell wall polysaccharides (glucans, in the case of yeast). Similar processes, often involving related proteins, can be identified in plant cells and even in crawling cells of animals. Perhaps the most instructive of these commonalities is that cell polarization takes place, not from the nucleus outward but (as Jaffe pointed out thirty years ago) from the periphery inward. A specialized cortical domain comes first, providing a spatial marker that links up with the nucleus. Polarization is the set of processes that establish cellular vectors for the flow of matter, energy and information.

AMOEBOID CELLS: FORMS IN FLUX

In everyday parlance the amoeba stands for all that is primordial and unformed; the very names *Amoeba proteus* and *Chaos chaos* imply fickleness and disorder. In reality, an amoeba is an efficient carnivore that manages to capture darting ciliates, and whose laborious crawl is not nearly as haphazard as it looks. Besides, the amoeboid way of life must be accounted an evolutionary success: innumerable amoebas of different kinds roam the soil (Fig. 7.1b), and amoeboid movements have been retained by many tissue cells of higher animals including our own leukocytes, fibroblasts, neurons and embryonic cells. Amoebas lack a rigid cell wall and do not maintain a fixed shape; the cell's form changes continually as it moves, grows and divides. Nonetheless, the various species each display a particular pattern of order that is recognizable under the microscope. An inquiry into amoebal forms should address both levels: how the changing form of an individual cell relates to its activities, and how the constancy of the underlying pattern comes about.

Research on amoeboid movements has generated an enormous literature covering both free-living amoebas and animal cells in culture dishes (14). We are increasingly well informed concerning the molecular basis of motility, much less so about the integration of molecular processes into coherent, functional activity on the cellular scale. As to the principles of design that govern the architecture of amoeboid cells in general and allow us to recognize amoebas and fibroblasts as variations on a common theme, almost nothing can yet be said. Efforts to model cells as tensegrity structures, describable in terms of continuum mechanics, promise a revival of interest in global principles of form and function.

The universal feature of amoeboid cells, as Grebecki (14) points out, is the absence of any motor oganelle endowed with stable position and permanent structure. In these cells, any activity leads in the end to the disintegration of the mechanical effector, which must thereafter be reconstructed in a new location. The point is illustrated by the life cycle of *Amoeba proteus* (Fig. 7.5), which traverses a succession of shapes ranging from vaguely cylindrical to nearly spherical. A foraging amoeba (Fig. 7.5a) is both functionally and anatomically polarized, though its outline is forever changing as the cell explores its surroundings with the aid of cytoplasmic protrusions called pseudopods. These can arise anywhere on the cell's surface except for the "tail," or uroid; one pseudopod that makes contact with the surface defines the main line of advance, the others are retracted. What looks like a drunken stagger becomes better directed when the amoeba senses prey and achieves a speed of several hundred micrometers per minute. When a full grown cell approaches division it retracts all its pseudopods and rounds up; a division furrow develops along the equator and splits the cell in two. Even before the process is completed, fresh small pseudopods emerge and help pull apart the two daughter cells (Fig. 7.5b-d). The entire cycle turns on reversible transformations of the physical state and mechanical properties of the cell cortex, which in turn are determined by a pervasive system of actin-based microfilaments.

Cyclic transformation of the state of the cell cortex also underlies the

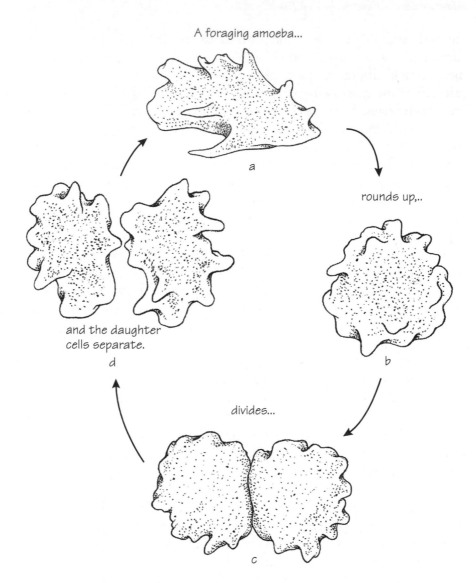

A foraging amoeba...

a

rounds up,..

b

divides...

c

and the daughter
cells separate.

d

Fig. 7.5. Scenes from the life-cycle of *Amoeba proteus*. A foraging, well fed amoeba (a) rounds up and retracts its pseudopods (b). Following equatorial division (c) pseudopods re-emerge (d) and pull the daughter cells apart. Length of amoeba 500 μm. After photomicrographs by Dr. David Prescott, with permission.

form and behavior of fibroblasts and other animal cells, though these differ greatly from amoebas in size, shape and habit (Fig. 7.6b). Here again morphology is in flux. Fibroblasts are fan-shaped and visibly polarized. A fibroblast advances by extending a broad, flat, and very thin pseudopod clear of organelles (called a lamellipodium), that attaches to the surface in front of the cell body. Needle-like spikes emerge along the cell margin, probe the space ahead and then retract. The dorsal surface commonly breaks into ruffles that drift towards the cell's tail, subsiding as they go. Sites of adhesion form on the ventral surface, spanning the plasma membrane and the cortex; some are linked to the perinuclear region by contractile cables called stress fibers. The distribution of these adhesion sites is a primary determinant of the fibroblast's shape. As the fibroblast creeps at a rate of millimeters per day, it periodically retracts the tail; the cell body moves up, and a fresh lamellipodium gropes forward. Fibroblasts, it seems, pull themselves by the front, whereas amoebas push from the rear. But the fibroblast, like the

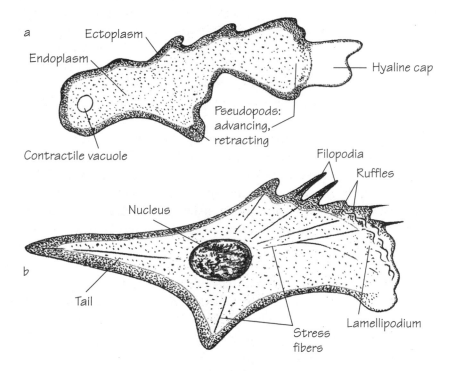

Fig. 7.6. Cortical structures in amoeboid cells. (a) An amoeba, emphasizing the distinction between the interior endoplasm and the ectoplasmic sleeve. Length, 200 μm. After Grebecki, 1994, and elsewhere. (b) A fibroblast with lamellipodium and stress fibers. Length, 70 μm

amoeba, both shapes and propels itself by continuously remodeling its cortical layer, whose mechanical properties are functions of an actin-based cytoskeleton.

The connection between cell movements and the thin, clear cortical layer that subtends the plasma membrane has been recognized for nearly a century. The cell cortex is comparatively stiff and strong, thanks in part to mechanical tension directed inward. But solid cortex is readily transformed into the more fluid cytoplasm that lies beneath. The conversion of rigid ectoplasm into mobile endoplasm (Fig. 7.6a), and back again, is especially conspicuous in the large amoebas. Physical chemists describe this as the interconversion of two states of a protein solution, called sol and gel (a homely instance is chicken broth fortified with gelatin, that solidifies into jelly but liquefies when warmed). In amoebas these two stages are segregated to opposite ends of the cell, and coupled by a circulation of cortical material from pole to pole. Gel turns into sol in the uroid; as the amoeba extends a pseudopod, fluid endoplasm flows through a peripheral sleeve of solid ectoplasm, sending nucleus and organelles tumbling forward (Fig. 7.6a). At the psedopod's front, ectoplasmic gel re-forms, to be dragged slowly toward the rear by mechanical tension. An amoeba's crawl is powered by a vectorial, energy-driven circulation of cortical material, which generates traction when the cortex is coupled to the substratum through foci of adhesion. We can see the overall process as another instance of the interplay of force and compliance.

The circulation of cortical material in fibroblasts and other small amoeboid cells is less prominent, but manifests itself in the movements of surface structures on the cell body. According to Bray and White (14), many kinds of cell movement ultimately reflect cortical flow. Cytokinesis (cell cleavage) is one more instance. Cortical material relaxes at the poles of a dividing cell, and is carried towards the equator by continuing tension in that region. At the equator, it assembles into the well-known contractile ring, that tightens progressively, not unlike a purse-string, constricting the cell's surface to make two.

From the molecular perspective, all this makes more and better sense with every passing year. The chemistry of the cell cortex revolves around the protein actin, a building material of almost infinite versatility, which makes up as much as a tenth of the protein complement of mobile cells. Actin monomers are readily assembled into polarized microfilaments, a chemical reaction governed and modulated by a whole tribe of auxiliary proteins. Actin-binding proteins modify the rates of monomer association and dissociation; they also render filament extension and shrinkage sensitive to such cellular parameters as the local acidity or the concentration of calcium ions, and to diverse regulatory metabolites. Actin-binding proteins govern the formation of higher-order complexes, and thus direct the mechanical behavior of the cell cortex. Fluid endoplasm contains chiefly actin monomers and short filaments, stiff ectoplasm consists of a meshwork of crosslinked microfilaments. Filopodia are stiffened by a core of microfilaments bundled together, stress fibers are cables of actin plus several other proteins arrayed into primitive muscle fibrils. The state of actin is intensely dynamic. Monomers come together into a crosslinked meshwork that then liquefies under the influence of proteins that sever actin filaments, cap their ends to block elongation, or take monomers out of circulation. These transformations, which often occur on a timescale of seconds, respond to local signals. Chemoattractants, for instance, direct the movements of slime-mold amoebas by eliciting localized actin assembly and therefore pseudopod protrusion. And in the giant amoebas, the vectorial circulation of cortical actin is governed by the local level of cytoplasmic calcium ions: high in the uroid, favoring solation, low at the pseudopod tip where gelation takes place.

To move a cell from its moorings, extend a pseudopod or drive a circulation of cortical proteins, energy must be harnessed and force must be produced. The underlying mechanisms have been debated for years, and consensus reached on two of them. The assembly of actin filaments and meshworks, both directional processes, can in themselves generate oriented force; this is presently thought to underlie protrusive activities such as the extension of filopodia and the initiation of pseudopods. The other force, which usually manifests itself as active contraction, is generated by sliding actin filaments against one another, a process mediated by myosin and powered by the hydrolysis of ATP. Contraction of actomyosin along the ectoplasmic sleeve and in the uroid is what propels the amoeba's cytoplasm. Biochemically speaking the process is the same that powers muscle contraction, but in amoeboid cells the spatial organization of the molecular components is less regular and more dynamic.

All the participants come in multiple versions which underpin the rich repertoire of forms, functions and behaviors. Let me single out one of these variations, because it throws into sharp relief the gap between molecules and the larger patterns of form and function. All creeping and crawling cells make use of actin, myosin, and a selection from the menu of auxiliary proteins, except for the amoeboid sperm of nematodes (roundworms, including the notorious tapeworm). These cells look and crawl like all the others. They extend pseudopods and form adhesion contacts, but they do it all with the aid of a cytoskeleton devoid of both actin and myosin (15). The instruments of morphogenesis and motility are filaments made up of "major sperm protein," a protein apparently unrelated to actin, which undergoes analogous polymerization and dissociation reactions. With myosin absent, and no evidence for contractile forces in sperm locomotion, polymerization of major sperm protein is thought to generate the driving force for pseudopod protrusion and motility.

Nematode sperm may be biological curiosities, but their peculiarities point to a larger moral. Clearly, form and function in amoeboid cells do not depend on a particular set of molecular players (though many details surely will). Any suite of molecules will do, so long as they can be articulated into cellular structures that support the function at hand. Form grows out of this organization of molecules, not their chemistry, just as a given building plan can often be executed in marble, brick or cement. It follows that the study of morphogenesis marches with engineering: its vocabulary includes terms that describe the properties of materials, patterns of force and of stress, supports and constraints, channels and flows. And always in the background, just out of earshot, a murmur of mystery: how are all these activities integrated into a pattern that works, reproduces itself and persists for millennia?

Arranging a Ciliate

One seeks to understand morphogenesis on two levels. One level, the more accessible one, centers on the mechanics of shape generation. This has been a productive line of inquiry; the preceding sections illustrate what we have learned concerning the forces that mold cells, and how cells respond to force in a controlled and localized fashion. We now turn to the deeper level, pattern formation; here we ask, not what cells do but how they direct what is done in space and in time. Spatial and temporal order, commonly intricate and highly reproducible, is the hallmark of every cell and organism. It leaps out at us, whether we are watching the living cell or pore over its ultrastructure, but how cells manage the choreography of their components is still nearly as baffling as it seemed to Alice Fulton two decades ago (16). Our ignorance in this matter constitutes a huge lacuna in our understanding of living things; it is fair to say that in the absence of satisfying ideas about pattern generation cells (and therefore life itself) remain fundamentally unintelligible.

A pattern, according to Webster, is simply "a coherent structure or design." Spatial patterns among single-celled organisms vary enormously, from the regular but plain designs of bacteria and fungal hyphae to the elaborate arrangement of cilia on the surfaces of protozoa. "Pattern formation," then, lumps together a universe of processes that arrange elements in space; it embraces spare elegance and baroque fantasies, the transformation of one pattern into another and the emergence of regularity from a prior state of disorder. Ciliates, among the largest and most complex single-celled organisms, highlight the problem and point the way towards a possible solution. The continuing effort to learn how ciliates position cortical organelles during growth, development and regeneration has provided incontrovertible evidence that such patterns are not spelled out in the genome. Instead, the placement of new cellular structures is directed by existing ones: structure begets structure. These researches have led to the rediscovery, at the cellular level, of morphogenetic fields—a concept that promises to illuminate the mystery of spatial patterning in general. The highly abbreviated account that follows draws on the extensive and thoughtful writings of Joseph Frankel, Gary Grimes and their associates (17).

Tetrahymena thermophila is large and complex for a cell, but simple as ciliates go (Figs. 7.1e, 7.7). The cells are pear-shaped, nearly 50 micrometers in length and ornamented with lengthwise, gently spiraling rows of cilia, which are responsible for its movements. At the cell's front is the mouth, equipped with specialized arrays of cilia whose beating drives food particles into the gullet; at the rear is the cytoproct, or anus, and a contractile vacuole for the expulsion of fluids, that are ejected via a pair of pores. Note that cortical organization is asymmetric; a normal cell, as in Fig. 7.7, with the contractile vacuole pores to the reader's left, is defined as right-handed. The relative positions of the mouth, ciliary rows and contractile vacuole pores are the chief landmarks that define the cortical pattern. Beneath the cortex lie two nuclei, a micronucleus reserved for reproduction, and an enlarged working macronucleus that supplies the huge cell with gene transcripts; these also occupy regular positions but will not be discussed here. The organism grows by extending its polar axis to generate two tandem segments, each of which duplicates the standard pattern. Therefore, before dividing at the "waist," the cell must construct a new contractile vacuole and cytoproct for the anterior daughter, and a new mouth for the posterior one (Fig. 7.7). Reproduction thus entails two general requirements: longitudinal extension of each ciliary row, and the production of specialized

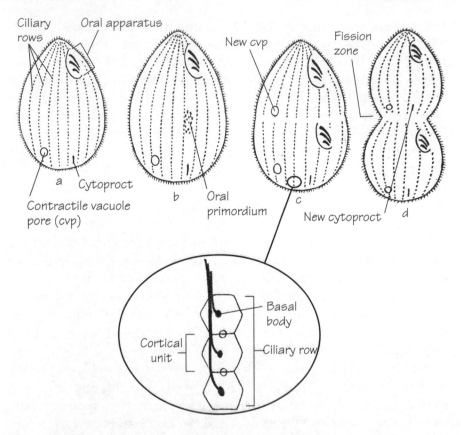

Fig. 7.7. Stages in the division of *Tetrahymena thermophila*. For description see text. Inset: geometry of a ciliary unit. After drawings by Joseph Frankel, with permisison.

organelles at sites remote from the original ones. The question here is, not how these organelles are assembled, but how the cell knows where to put them.

It may be helpful to the harried reader to anticipate how far we have come in the search for an answer. Four decades of research have confirmed and amplified the late Tracy Sonneborn's original insight (18), that new cell structures are ordered and arranged under the influence of preexisting cell structures. The informational system that directs the assembly of organelles in space and time is complementary to, but separate from, the instructions encoded in the genes. We infer this from the observation that spatial patterns are commonly inherited by pathways that do not involve genes, but turn on the structural organization of the whole cell. We do not yet understand the physical and chemical mechanisms that underlie directed assembly (to borrow a phrase form G. W. Grimes), but it is clear that these fall into two classes: local mechanisms, on the scale of molecules and large complexes, and global ones that extend over the entire cell. The properties of the latter recall what embryologists have traditionally called a field: "a territory within which developmental decisions are subject to a common set of coordinating influences" (17).

The local level of directed assembly, sometimes referred to as structural guidance, is best illustrated by the propagation of ciliary rows, which was clarified thirty years ago by Sonneborn and his colleagues; they worked with another ciliate, *Paramecium*, but the conclusions apply just as well to *Tetrahymena* (Fig. 7.7, inset). A row consists of a linear array of ciliary units, each endowed with its own local geometry. At the center is a basal body from which the cilium springs; a striated rootlet and the accessory band of microtubules extend toward the cell's anterior right. Each ciliary unit thus has an intrinsic asymmetry, and the beating pattern is determined by its geometry. Rows propagate by the insertion of new basal bodies anterior to existing ones, and with the same orientation; they acquire accessory fibers in conformity with those already present, and finally sprout cilia that beat in unison with others in the same row. Occasionally part of a row becomes inverted, either by accident or as a result of experimental manipulation. Inverted rows propagate just as normal ones do: they maintain their intrinsic polarity, but new ciliary units are now inserted on the side facing the cell's rear, and all accessory structures are likewise rotated by 180 degrees. "The site of initiation of basal body assembly, its path of migration to the surface of the cell, and the organization of associated structures around it are [all] determined by the molecular geography within the unit territory, and not by any outside influence, either nuclear or cellular" (18).

We do not presently understand the mechanisms by which existing structures constrain or guide the assembly of future structures. Molecular self-assembly (as in the spontaneous association of proteins to make a simple virus or a ribosome) is surely involved, but something more like a template seems to be called for. What we do know is that structural guidance is a common mode of pattern formation, at least among ciliates, for there are many examples of modifications that are transmitted from one generation to the next, not by way of the genes but because they are linked to a cellular structure that persists through cell duplication. For example, contractile vacuole pores in *Tetrahymena* are located posteriorly and to the right of a particular ciliary row; should that row be inverted, the pores appear on its left, presumably because the vacuole is connected to the ciliature. The most spectacular instance of the persistence of cell structure is the propagation of doublet cells, back-to-back fusions that occasionally arise as a result of abortive division or of surgical manipulation. Doublet cells can propagate indefinitely as doublets by a mechanism that depends, not on gene mutations but on the continuity of surface organization from one generation to the next. Some feature of cortical organization, apparently something other than the ciliature itself, carries over and governs the configuration of the entire cell cortex; one would dearly like to know what this persistent skeleton may be.

Can all instances of patterning be attributed to direct and short-range interactions among specific parts? The answer is definitely no, and the discovery that global patterns are independent of (and superimposed upon) local structural guidance must be one of the signal contributions of ciliate research to general cell biology. The influences that generate global patterns have properties traditionally associated with developmental fields: the positions of cellular organelles are not fixed, but assigned relative to whole-cell parameters, the pattern is the same regardless of the cell's size, and in many cases a part of the pattern can reconstitute the whole. The evidence has been marshalled by Frankel and by Grimes (17), and there is no way to do it justice here. Suffice it to indicate in a general way what distinguishes global patterns from the local variety.

There are now several instances in which the local pattern is reversed while the global one is normal. Inversion of a ciliary row rotates, not only the orientation of new ciliary units but also (as mentioned above) the placement of the contractile vacuole pore with respect to that row. However, the pore continues to develop in the posterior half, and selection of the particular rows that bear the pores is likewise unaffected by rotation of the rows; these are matters for a global assessment. Conversely, there are situations in which the global pattern is altered while local ones remain unchanged. A spectacular case in point comes from the reversal of cellular "handedness." Normal cells of *Tetrahymena* are "right-handed" (Fig. 7.7), but "left-handed" ones arise under certain conditions and can propagate their peculiar asymmetry. In left-handed cells, cortical organelles are wound around the circumference in the opposite sense; the oral apparatus, which in the normal cell, illustrated in Fig. 7.7, faces to the reader's right, instead faces left. Yet the ciliary rows and ciliary units of the two cell kinds are superimposable! It appears that ciliary units can exist only in a single isomer, the one illustrated in the inset, whereas the global axis can flip right over to specify a mirror-image configuration.

That the global positioning system has the organizational character of an embryonic field is inferred chiefly from studies on the larger ciliates. The capacity of members of the genera *Stentor* and *Oxytricha* to recover from physical insults almost exceeds belief (and assuredly passes understanding). In now classical research carried out in the fifties and sixties, Vance Tartar (a recluse and an original, such as would scarcely survive in today's science, ref. 19) found that starving *Stentor* cells undergo successive reorganizations, each of which results in a smaller but quite normal cell. The organs (e.g., the mouth) each become smaller, but the pattern (the spatial relationships between the parts) is preserved down to a lower limit, beyond which a viable cell can no longer be constructed. Tartar also discovered the legendary capacity of cell fragments, some as small as 1/100 of the whole cell, to reconstitute a tiny but complete and viable *Stentor*. Regeneration requires that the fragment contain a part of the macronucleus and a particular region of the cell cortex; the latter serves as the organizing center that directs the reconstitution of the oral primordium and all other cortical structures.

How can we imagine a unicellular organism knowing how to position the ciliary arrays that make up its oral apparatus, let alone reconstitute the whole pattern from a fragment? In truth, we have no more than a glimmer of an idea, which turns on the proposition that global positions are specified by coordinates rather like the familiar latitudes and longitudes. Positional information along the equatorial meridian would be read out along a circumferential gradient that specifies such landmarks as ciliary rows and contractile vacuole pores. A gradient from front to rear, at right angle to the first, locates the oral apparatus and the posterior organelles. The equatorial gradient propagates continuously as the cell elongates, which accounts for its persistence; the antero-posterior one, which must be reorganized every time the cell divides, is much more labile. The particular merit of this idea, which has been developed rigorously and in detail by Frankel (17), is its capacity to account for the peculiarities of certain mutants and surgical constructs in which the continuity of positional values appears to be disrupted.

It is a virtue of the cylindrical coordinate model of pattern formation that it is altogether abstract; its usefulness stems from a rule that states that positional values must be continuous at all times, like the numbers on a clockface. If surgery or mutation perturb the continuity (putting a four next to a two, in place of a one or a three), the organism takes measures to restore it, sometimes with bizarre developmental results. The validity of such a model can be examined, up to a point, even in the absence of information about the materials that make up the grid, but its refinement surely requires one to identify the entities and forces that underlie the formal rules. To the visible frustration of investigators in this area, progress along that line continues to be stymied. Electron microscopy has not revealed anything that could correspond to a grid of positional values, biophysical approaches have not identified plausible gradients of mechanical or electrical parameters, and as yet we do not know in biochemical terms what ails the mutants in which the specification of pattern is perturbed. There is no shortage of ideas, some of which will be outlined in the following section, and optimists expect the crucial clue to pop up momentarily. But it may also turn out that we are in the position of the Mullah Nasruddin, who dropped a gold coin in the bazaar but insisted on searching for it under the lamp because the light was better there.

Perhaps we should prize the doubt, for the very intractability of the problem assures us that we are confronting a genuine mystery. One can argue that ciliates are unique creatures that have invented developmental mechanisms without parallels elsewhere. For myself, I suspect that ciliates are not extraordinary, except insofar as their elaborate cortical architecture parades a capacity for large-scale pattern formation that is universal but generally inconspicuous. After all, spatial order on the scale of micrometers to millimeters is visible in many cells, and we have already taken note of the widespread use of spatial markers in cell morphogenesis and continuity. When we do at last learn how these cells position their oral ciliature or contractile vacuole pores, the gains are likely to reach far beyond the humble organisms in which the discovery was made.

IN PURSUIT OF WHOLENESS

Morphogenesis illustrates at the cellular level what Warren Weaver meant when, fifty years ago, he identified the problems of organized complexity as biology's high frontier. A cell of *Tetrahymena*, say, with its particular morphology, anatomy and life cycle represents a pattern in space and time. Each such pattern coordinates the activities of innumerable molecules into a unified structure, reproduces itself periodically and persists in this manner indefinitely. From the examples discussed in the preceding sections it seems self-evident that the generation and perpetuation of such extended patterns cannot be understood solely in terms of the local, random, and scalar chemical events that are the stuff of molecular science, but depends on organizing principles that operate on the scale of cells and organisms three to five orders of magnitude larger. Curiously, this perception seems quite foreign to the majority of experimental biologists; its primary home is the still-emerging field of complexity studies, inhabited by physicists, mathematicians, and computer mavens (20). Whether ideas spawned on those marches of experimental science can reveal principles of pattern formation and morphogenesis that have eluded us so far still remains to be seen. The musings that follow draw on the writings of many of the pioneers, particularly on those of Lionel Harrison and Brian Goodwin (21), whose influence pervades this section.

Consider for a moment how, as a matter of physical principle, regular forms and patterns arise in nature without the intervention of living organisms. There are not many. One way turns on structure: molecules come together into crystals according to rules inherent in their geometry and chemistry. In like manner, macromolecules assemble more or less spontaneously into ribosomes, viruses and membranes. Many of those who approach living organisms from the genetic and molecular perspective appear to believe that such molecular agglomeration can, in principle, account for the form of cells, even organisms. I have argued earlier (Chapter 5) that local interactions will not suffice because a cell is not a self-assembling structure, and a little reflection on the preceding examples of eukaryotic growth and morphogenesis should resolve any lingering doubts. The other purely physical sources of spatial order turn on kinetics: many dynamic physical systems organize themselves spontaneously into extended patterns whose basis resides in flow rather than structure. One such is the Bènard instability, the spontaneous emergence of discrete convection cells in a pan of oil uniformly heated from beneath. Others are a flame, a thunderstorm and a whirlpool. Chemists are particularly fond of the Beloussov-Zhabotinski reaction, a mixture of certain organic and inorganic chemicals that generates concentric rings of colored products that diffuse outward over a distance of centimeters. Biologists have never been persuaded that such purely physical and chemical processes are relevant to what goes on in an amoeba or an embryo, because we are so firmly wedded to the genetic program that informs living organisms, but is absent from nonliving ones. But dynamic patterns generated by physical and biological systems do have something in common that hinges on the concept of a *field*.

Like other metaphors imported from everyday life, a field is an elastic idea whose content depends on the speaker. I shall employ the term in a very general sense, to designate a territory that displays coordinated activity controlled by the differential distribution of some property or agent. The virtue of this abstract notion is that it lends itself to mathematical formulations that incorporate such features as continuity of field values at every point in space, smooth transitions and directional change. Fields have the holistic quality that, given a global mathematical expression and a few local numerical values, it is often possible to reconstruct the field in its entirety. Furthermore, since the essence of a field resides in its mathematical description, one can examine the properties of a field without knowing anything about its physical nature. That is a great advantage, for the agents and properties whose distribution determines field behavior come in many forms. Fields of force (electrical, magnetic, or gravitational) are familiar, but fields of biological interest can also be sustained by a concentration gradient, or by a pattern of mechanical stress and strain. The fields most pertinent to morphogenesis and patterning are those generated by dynamic rather than static systems; the flame-like character of living things is more than a poetic simile.

Dynamic systems are characteristically maintained in a state remote from equilibrium by a continuous flow of energy. Given the right parameters, physical systems of this kind commonly undergo spatial self-organization, with concurrent enhancement of the energy throughput (when that heated pan of oil produces convection cells, the rate of heat transfer rises). Such patterns were designated "dissipative structures" by Ilya Prigogine, who regards them as one of the chief sources of order in the universe (20). Note that, like a living organism, a dissipative structure coordinates the random motions of innumerable particles over an extended territory, and may persist indefinitely so long as the supply of matter and energy lasts. The behavior of dynamic systems is typically non-linear. Over a certain range, an incremental input of energy or matter produces an incremental output, but at a particular threshold there is an abrupt change in behavior (Bénard's pan of oil or an excitable cell). Non-linearity is commonly a consequence of feedback interactions among coupled processes; their mathematical description calls for a sequence of coupled differential equations. To be sure, designating a growing hypha or a regenerating ciliate as a dynamic field does not in itself explain anything. But the label helps to focus the mind on the features that call for explanation, and it highlights parallels with the physical world that can be described with a common formalism. The fact that, in a growing number of instances, the field formalism rationalizes or predicts biological behavior, and sometimes allows one to compute the shape an organism ought to display, engenders confidence that there is more to this than formalism alone.

The general proposition that pattern formation and morphogenesis are directed by a dynamic physical field is far from novel. Embryologists have thought along these lines since the twenties, and an explicit hypothesis was set out nearly fifty years ago by Alan Turing as a solution to a somewhat different question: How can spatial order arise from a prior state of disorder, as happens when an apparently homogenous egg turns into an embryo? In a paper boldly entitled "The chemical basis of morphogenesis," Turing (22) put forward two important ideas. One was the proposal that developmental events are called forth by specialized informational molecules, called morphogens, whose distribution in space supplies a prepattern for the subsequent location of biological structures. In contemporary idiom, the graded distribution of morphogens constitutes a field of positional information, a kind of map, that instructs individual cells in an early embryo concerning the developmental course that each should follow. The other, and entirely novel, idea was that a pattern of local concentration differences can arise spontaneously when two interacting substances diffuse at different rates. Contrary to intuition, which associates diffusion with smoothing out concentration differences, in systems that obey particular kinetic rules random fluctuations arising within a homogenous region will be amplified, generating stable local maxima and minima of morphogen concentration. Note that Turing's principle is grounded in physical chemistry; he worked out its biological implications in terms of multicellular embryos, but there is nothing to preclude its extension to single cells.

The general hypothesis, that a pattern of morphogen distribution guides biological development, is very much alive. It has proven directly applicable to the development of animal embryos, in which the major axes are blocked out by gradients of diffusible substances (proteins, as a rule) that instruct cells concerning their position in the embryo and direct them into the proper path of differentiation (23). Clear examples come from embryonic development in the fruit fly, *Drosophila*. Early in this process, a gradient of a protein called Bicoid arises by the localized translation of mRNA which had been deposited by the mother fly in the region destined to become the embryo's head. At this stage the embryo contains many nuclei, but these are not yet separated by cell membranes. The Bicoid protein activates the expression of certain genes involved in the establishment of the fly's segments, and nuclei respond differentially according to the concentration of Bicoid they encounter. Those at the high end of the gradient are induced to embark on the production of the head, lower Bicoid concentrations induce other anterior organs. Additional morphogens specify posterior structures and the differentiation of the dorso-ventral axis.

Can this particular model be applied directly to the specification of pattern and form in single cells? Probably not, for several reasons (but see Harrison, 21, for an alternative view). First, a unicellular organism has no population of nuclei that can respond differentially to the local concentration of some instructive substance; very different mechanisms would be required for a morphogen to direct, say, localized exocytosis or the disposition of cilia. Second, the cytoplasm of most eukaryotic cells (including ciliates and fungal hyphae) is constantly stirred by streaming; local concentration differences of diffusible morphogens would soon be erased, unless they are confined somehow to the quiescent cortex. Finally, there is an awkward dichotomy between a map of morphogens that carry instructions and the separate interpretation of those instructions. Such a division of labor seems plausible on the scale of a fly embryo (half a millimeter in length), but not on that of an individual cell. Still, we should hold on to the general proposition that form and pattern on the cellular level are the expression of a field over which some agency acts in a coordinated manner. This agency might be electrical in nature, or a pattern of mechanical stress and strain, or something else altogether; and it need not be the same in all cells. Beyond laws that apply to dynamic fields generally, there may not be much unity beneath the quirky diversity of biological forms.

Fields are abstractions; to make them concrete, the general concept must be applied to particular cases by specifying the informational properties of the field, how it arises, and how its instructions are implemented. The most persuasive cases are those in which one can compute the pattern or form of an organism from a set of explicit premises. Once again, the fact that some formula or algorithm generates a biological shape does not guarantee that the mathematics capture the underlying physiology; but success does indicate a well-crafted hypothesis whose postulates can be verified or challenged by more empirical methods.

The placement of cortical organelles in *Tetrahymena* illustrates at once the powers and limitations of the field approach. In the model developed by Brandts and her colleagues (24), the central postulate is that the cell's surface can be represented as a smooth and continuous set of positional values arrayed around the circumference. Various organelles, such as the oral apparatus and contractile vacuole pores, correspond to particular values; the model does not specify whether these field values merely designate map positions or are instrumental in the construction of organelles. The field as a whole has a single value, designated an "energy," which is calculated by summing the energy density at all points. This energy is made up of two terms, one capturing the idea that there is an optimal gradient or spacing of cortical features, the other that changes in the magnitude or direction of the gradient cost energy.

The cell seeks configurations that minimize the overall field energy by adjusting the two terms. The output of the model consists of a set of patterns of oral apparatuses and contractile vacuole pores, that are allowed by the formulation and vary as a function of the cell's circumference. Doublets, singlets and transitional intermediates should exhibit different configurations.

This is an exceedingly abstruse model, unlikely to commend itself to either biochemists or physiologists. Yet it makes remarkably accurate predictions concerning the configurations of organelles to be expected in individual cells and populations. It predicts that reversion of right-handed doublet cells to the singlet state should follow a path different from that of left-handed doublets, and correctly forecasts the intercalation of a third oral apparatus between the other two. It also calls for certain configurations of contractile vacuole pores that were observed only after the model had anticipated their existence. This predictive power suggests that the model captures essential aspects of biological reality, and highlights the requirements that must be met by any system of real molecules specifying positions in real cellular space. Such a system may consist of two or more species of mobile molecules, diffusing Turing-fashion within the cortical layer and interacting in such a way that the positional interpretation reflects the ratio between the two species. But it seems more likely that the molecular basis of the positional field should be sought in physical parameters of the cortical layer itself. The game is afoot, and that the quarry continues to elude its pursuers should lend zest to the hunt.

For Brian Goodwin and Lionel Harrison, the object of study is growth and regeneration of apical structures in the giant unicellular marine alga *Acetabularia* (Fig. 7.1f). Briefly, when the umbrella-shaped cap is lopped off the stem forms a tip, elongates and eventually regenerates the cap; in the process it also puts out successive whorls of hairs that have no known function and soon fall off. Goodwin, Trainor and their associates set out to devise a mathematically explicit model grounded in established cell physics that specifically predicts this sequence of events; and while that goal has been achieved only in part, a very good start has been made (25). They begin by considering the stem as a closed vessel consisting of three apposed elements: a stiff and strong wall, a fluid-filled vacuole that exerts hydrostatic pressure (turgor) upon the wall, and a thin shell of cytoplasm sandwiched between the vacuole and the wall. Pattern is generated within the cytoplasmic layer thanks to reciprocal interactions between the concentration of free calcium ions and the mechanical state (stress and strain) of the cortical cytoskeleton.

The emerging pattern is transmitted to the wall, perhaps by the intervention of ion pumps in the plasma membrane, and this localizes the wall's expansion in compliance with the force of turgor. All this and much more was incorporated into a set of coupled differential equations, more than twenty of them, which define a morphogenetic field within the apical region. The physical nature of this field is very different from that envisaged by Turing, but their mathematical properties turn out to be similar.

What makes this abstract model a serious contribution to the science of form is its capacity to generate realistic shapes from a nearly uniform initial state. Computer simulations begin with the apex as a low, featureless dome. As the program goes through its paces the dome puts forth a tip that advances and then flattens, just as the real tip does. The apical calcium concentration peaks at the apex, then turns into an annulus that breaks spontaneously into a series of peaks that have the symmetry of a whorl of lateral hairs. Harrison's laboratory demonstrated earlier that calcium ions are involved in hair production, and that calcium accumulates at the site of hair emergence. All this suggests that the lateral hairs do not represent a functional structure, but emerge as a consequence of system dynamics. It has not been possible to model the emergence of the hairs themselves, or of the reproductive cap, partly for technical reasons and partly because even this very complex model contains far fewer terms than the living alga employs.

Note that this model does more than describe shapes; it generates them thanks to its internal dynamics, just as the living cell must do. One might expect this gratifying outcome to be critically dependent on the numerical values assigned to the many parameters that must be specified, but that is not the case. The model is robust, in the sense that it "works" over a broad range of parameter values, generating a family of forms and sequential transformations. The biological implication is that the morphology of a regenerating tip is both stable and probable, a form that will emerge naturally in diverse cellular systems. Indeed, apical extension is a widespread mode of growth. *Acetabularia* itself is one of a large order of related algae, the *Dasycladales*, whose fossil remains go back to the Cambrian era 500 million years ago. Goodwin does not doubt that these algae are related historically, by descent from a common ancestor, but the fact that they display variations on a common morphological theme is explained, not by their common ancestry but by their shared physiological dynamics. "From this perspective, the *Dasycladales* constitute a natural group not because of their history but because of the way their basic structure is generated" (25).

To put it in technical lingo, they constitute a discrete basin of attraction in morphospace. Goodwin also sees a conflict between such "laws of form," rooted in systems dynamics, and the common understanding that natural selection has shaped organisms over time; but here I must part company from him. We shall revert to this subject in Chapter 9.

Let me conclude this section by returning briefly to a somewhat simpler example of cellular morphogenesis, apical extension in fungal hyphae, which was discussed in some detail above. It will be recalled that the tip of a growing hypha contains a prominent vesicular structure known as the Spitzenkörper, or apical body; observations suggest that this body plays an important but ill-defined role in the passage of secretory vesicles to the site of exocytosis at the extreme tip. Salomon Bartnicki-Garcia and his colleagues (9) succeeded in modeling hyphal extension on the premises that the Spitzenkörper is the immediate source of secretory vesicles which are discharged at random at a rate N, and that this body is endowed with directional mobility, traveling at a rate V. Vesicles that reach the surface are incorporated, causing it to enlarge; and since vesicles shot out in the direction of travel will reach the edge soonest, the surface will preferentially expand ahead of the Spitzenkörper. Now, this is surely a brutally simplified description of hyphal extension, but it could be expressed in an extremely simple equation with N and V as the sole variables. This plots out as a curve, dubbed a hyphoid, that is instantly recognizable as that of a fungal hypha in cross-section.

On a previous occasion, I have presented a critical analysis of this hypothesis and of the evidence that is beginning to weigh in its favor (26). Here we will only note that this is once again a field theory: the advancing Spitzenkörper spraying vesicles in all directions generates a field in which exocytosis takes place in a coordinated and predictable manner. The model developed by Bartnicki-Garcia looks much simpler than Goodwin's because most of the physiological complexity of hyphal growth is subsumed under the collective variables V and N. This drastic summation allows one to formulate a model that is comprehensible and experimentally testable; and the reports from the laboratory suggest that it captures something fundamental to the way hyphae grow and shape themselves. Incidentally, the proposal that calcium influx localizes the apex represents yet another instance of a spatially extended field, one that has not yet been expressed in mathematical language.

When the field concept is applied to apical growth in fungal hyphae, it loses something of the precision that it brings to the specification of map coordinates in a ciliate. But that is not altogether detrimental, for the spatial organization of a whole eukaryotic cell is not likely to be dictated in a straightforward manner by a single master gradient. The major features of hyphal organization probably stem secondarily from the vectorial extension of the cytoskeleton, with its attendant traffic in vesicles and localized exocytosis. These establish the polarized organization of the plasma membrane and localize the origins of subsidiary gradients, including gradients of cytosolic pH, of membrane transport proteins, of wall mechanics and signal-carrying molecules. One thus arrives at the notion of a hypha as an extended matrix of multiple interwoven gradients, all of which are ultimately consequences of vectorial tip growth, and many of which also feed back upon tip extension. As in the case of parallel processing by informational networks, redundancy is built into the system. Multiple interactions may prevail over rigid hierarchy, and there may be no one indispensable vector; linear causality then dissolves into a web.

Biologists are apt to be uncomfortable with field theories, and understandably so. As matters stand there is still something half-baked about their application to biological organization, and in any event you cannot isolate a field, clone it or patent it. But it seems that we need some such idea, if only to rationalize the paradoxical relationship between morphology and genes. Everyone knows that the forms of cells and organisms are quite strictly inherited, and can be altered by mutation with specific and reproducible effects. Yet forms cannot be explicitly engraved in the genome; they are remote implications of the genetic instructions, each arising by the collaboration of numerous gene products distributed in space (Chapters 5, 6). I take the position, argued compellingly and in detail by Goodwin and others (27), that a spatially extended dynamic field generated by the cell as a whole is an obligatory intermediate between genes and form. Its function is to organize gene action in space: the morphogenetic field is the agency that defines the pathways of molecular transport and positioning, and ultimately localizes the forces and compliances that shape the cell. Fields remain hypothetical, and their physical nature a subject for speculation and research; but it seems to me self-evident that morphogenetic fields must revolve around the organization of the cytoskeleton. The particular field (or more likely, fields) that guides morphogenesis need not be the same in all organisms, but organisms related by descent will surely share field dynamics, just as they share gene sequences and molecular architecture. With each generation, the morphogenetic field is recreated afresh. The reason that forms are nevertheless faithfully transmitted is that each cell carries two kinds of heritable information: the linear sort, written in nucleotide sequences, and the three-dimensional sort embodied in the spatial architecture of the cell as a whole. Genes specify macromolecular functions, and collectively determine the kinetic and thermodynamic parameters of the morphogenetic field. The structural markers and cortical domains that are turning up in the molecular descriptions of cell morphogenesis are part of the mechanisms that define field boundaries, and fields defined by vectorial physiological process ultimately shape the cell. I realize, of course, that this point of view is thoroughly out of fashion; Jan Sapp (28) recently referred to those who hold it as "cytoplasmic heretics." But there is really nothing radical about it, or even novel; it's largely common sense, and therefore questions about the existence and nature of morphogenetic fields ought to rank high on the research agenda.

What, then, of the relationship of form to function: are cell forms shaped by the predictable workings of a physical system, or selected gene by gene for superior function? Both, surely, and unlike Goodwin I see no necessary conflict between these two viewpoints. Systems behavior determines the organism's form, which is in most cases plainly adaptive. There may well be morphological features that are not the result of direct selection, such as those whorls of lateral hairs upon a stem of *Acetabularia*, or the wing patterns of the desmid algae. But the fields that apparently guide morphogenesis should themselves be products of variation and selection. Most, though perhaps not all, of the variation that affects field dynamics occurs initially at the gene level; and most, though possibly not all, of the winnowing that modifies and transforms morphological patterns actually judges whole organisms by their functional performance. And when one reflects upon fungal hyphae forging across an agar plate, or a hungry amoeba trapping its prey with ponderous dexterity, form and function become indistinguishable.

8

THE ADVANCE OF THE MICROBES

"The reason for trying to understand the universe isn't that we thereby blunder into a new material for coating non stick frying pans. It's that we gain an insight into our place in the scheme of things, and of just how wonderful and unexpected that scheme can be."

Ian Stewart and Martin Golubitsky (1)

A PAIR OF DISTANT MIRRORS
THE ULTIMATE ANCESTOR
A MOST PECULIAR ALLIANCE
THE RISE OF THE EUKARYOTES

I love natural history museums, and best of all that grande dame on New York City's Central Park West, the American Museum of Natural History. Fifty years ago, its dim and dowdy halls furnished a comfort zone for this young immigrant, unsure of himself and of this vast strange land full of opportunities and terrors. The museum's new incarnation, splendidly refurbished, still feels like home. Its exhibits kindled an early interest in life's history, and my understanding of evolution is drawn largely from the writings of scholars attached to this museum and others elsewhere. I regret only that more than three quarters of that history does not lend itself to public display, for it is wholly the record of microbial life. (Fig. 8.1)

Multicellular creatures large enough to leave showy fossils first appear quite late in the game, about 600 million years ago, with the enigmatic Ediacaran fauna; organisms clearly ancestral to contemporary ones come later still, about 540 million years ago. By then, the most creative period of cellular evolution was long over; the fundamental patterns of metabolism, heredity and structural organization were all fixed during that

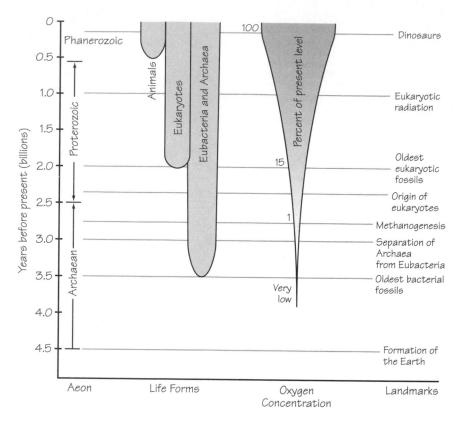

Fig. 8.1. Timeline for the history of life.

vast span of three billion years when the earth was populated exclusively by microorganisms. The object of this chapter and the next is to survey what has been learned concerning the emergence and proliferation of those cellular archetypes, and to listen for echoes of the underlying causes and forces. But the reader be warned: this inquiry is as much an exercise in myth-making as in biological historiography, for conclusions about the early evolution of living order must be drawn most gingerly from a fragmentary geological record and from the cryptic molecular files preserved in the genomes of contemporary organisms.

Historically, our conception of evolution has been based predominantly on the study of animals and plants, living and extinct. By contrast, evolutionary studies in deep time rely primarily on the historical record embedded in the sequences of contemporary nucleic acids and proteins. A prodigious volume of such information is at hand, and it affords biologists their "first glimpse of the full evolutionary landscape" (2); one in which microorganisms rather than dinosaurs fill the horizon.

The cornerstone for all current research and reflection in this field is the universal tree of life, based on the comparison of ribosomal RNAs from hundreds of organisms (Chapter 3); a simplified version is shown in Fig. 8.2. In a highly abstract manner, the tree encapsulates the whole history of life. It displays three great stems, two prokaryotic and one eukaryotic, that diverge from a common ancestor early in evolution and remain separate thereafter. The earliest divergence (the "root" of the tree) divided the primordial prokaryotic world in two; Eukarya, which arose somewhat later, are distantly but specifically related to the Archaea. Contemporary organisms represent the tips of branches from the central stems, some very ancient and others relatively recent; unfortunately there is no simple relationship between evolutionary distance and the passage of time, and therefore the universal tree has no intrinsic time scale.

Like most scientific diagrams, Fig. 8.2 represents both a summary of the data and their interpretation, and it implies more than we know. The very act of connecting points by lines lends an air of authority to what is at bottom a bold theory. Just in the past few years, new findings

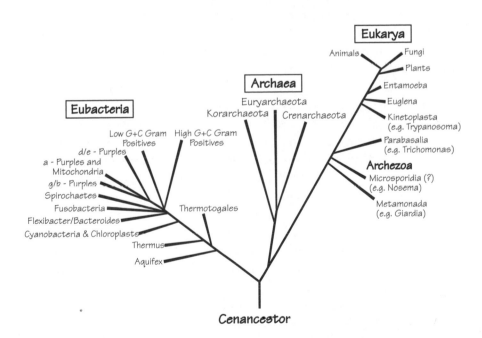

Fig. 8.2. A universal tree of life. Phylogenetic relationships inferred from ribosomal RNA sequences. The lengths of the lines are proportional to evolutionary distance, not time. The position of the tree's root is approximate. From Brown and Doolittle, 1997, with permission of the American Society for Microbiology.

have appeared that conflict with the universal tree; not all students of microbial evolution would subscribe to it, and some would redraw the scheme altogether. In my judgement, radical revision is unwarranted. As we proceed, we shall find reason to elaborate and amend the universal tree, but none to fell it and start over. On the contrary, it seems to me important to begin this chapter by underscoring the enormous intellectual achievement that the universal tree represents. For the first time, thanks very largely to Carl Woese and his disciples, we have a global and objective framework for reflection on how the living world came to be as we find it. The great tree is likely to be seen as one of the triumphs of biology in the twentieth century.

For an organismic biologist, the most significant episode in that span of some 3 billion years (second only to life's origin) is surely the advent of eukaryotic cells. This proposition does not leap out from the universal tree of life, which shows the eukaryotic branch just a little longer than the other two. Recall that those branches report the divergence of ribosomal RNA sequences, and while that is a most excellent way of tracing lines of descent it understates degrees of difference that emerge at the level of cellular rather than molecular organization. An interstellar naturalist collecting specimens from the Proterozoic ocean could not have foreseen what was to come, but those occasional large and florid cells caught in the sample were pregnant with infinite possibilities. Multicellular bodies, complex shapes and behavior, followed in due course by consciousness and rational thought—all these fall beyond the bounds of a prokaryotic world. For that reason I am not content with Gould's assertion that we still live in the age of bacteria, "as it was in the beginning, is now and ever shall be" (3). Judging by their biomass and by their contribution to the planetary economy that may well be true, but it is the eukaryotes that point the direction to increasing diversity of form and function. Biologists are reluctant, and rightly so, to describe evolutionary history as progressive, a term laden with connotations moral, religious and political. All the same, one comes away from a stroll through the fossil gallery with an overwhelming sense of mounting complexity and sophistication. Whether this trend is a real feature of evolution or only apparent, and if the former, what its nature and causes may be, are presently matters for debate. But if there is any larger meaning to the history of life, the rise of the eukaryotes marks a distinct stage.

Cosmic potential apart, the prokaryotic and eukaryotic tracks look quite different. The trademark of the prokaryotes is metabolic diversity. They discovered all the sources of energy utilized by contemporary organisms, developed the requisite molecular machineries, and still boast the greatest variety of economies. Prokaryotes are ubiquitous in soil, sea and atmosphere, in the earth's rocky crust, and inside other living organisms, and they occur in huge numbers wherever they are found. They are also small, single-celled and their elementary shapes have changed little over the past three billion years. To be sure, it's the exceptions that keep one honest: the gigantic bacterium *Epulopiscium fishelsonii*, half a millimeter in length, which roams the intestines of certain tropical fishes, or the shapely fruiting bodies of myxobacteria, each the social enterprise of millions of individual cells that come together for the purpose of reproduction. All the same, these exceptions probe the rule without overturning it. Eukaryotic cells, by contrast, introduced just a handful of biochemical innovations (sterols for one) and were otherwise content to embroider variations on the biochemical themes pioneered by prokaryotes. Where eukaryotes excel, and the "lower" ones in particular, is in the diversity of shapes, lifestyles and adaptations, their specialty is organization. The endless variety and immense range, from protozoan to philosopher, delights the eye even as it challenges the mind to explain how it all came about and what it means.

Students of cell evolution labor under a grave handicap that has no obvious remedy. Molecular technology, its powers growing by the day, lets one trace the lineage of particular genes and the macromolecules that these encode. On the premise that the phylogeny of genes tracks that of organisms (a proposition that, until just the other day, was taken to be self-evident), molecular phylogeny can supplement the geologist's meager gleanings. But physiology, the integrated functions that underlie life, leaves few traces in either genes or rocks, and this is the level most pertinent to those who would understand how the order of life came to be as we find it. We must perforce rely on what we can learn from contemporary organisms, and extrapolate into the remote past. Perhaps for that reason, one senses in evolutionary biology a freedom and a playfulness that have long since been extirpated from more exact sciences. "How enjoyable, how very enjoyable and luxurious it is, to suddenly emerge from the stern labyrinth of fact onto these dawn-lit uplands of surmise" (4).

A Pair of Distant Mirrors

Our mental image of bacteria as a class of organisms was shaped half a century ago, in terms of a fundamental dichotomy between relatively simple prokaryotic cells and the more elaborate eukaryotic ones

(Chapter 3). Bacteria, and *E. coli* in particular, became the organisms of choice for intensive biochemical scrutiny, which laid the foundations for the triumphalist molecular science of the present day. But progress in the study of bacteria as organisms was seriously hampered by the lack of a rational way to classify them and to trace their evolution. Attempts to derive bacterial taxonomy from their morphology and metabolism generated much confusion, and one microbiologist of the time wistfully likened the bacterial world to a great tree shrouded in dense fog showing only the tips of its branches.

This frustrating situation was radically transformed by Carl Woese's penetrating insight that a satisfactory taxonomy and phylogeny of bacteria can only be derived from the digital sequences of their macromolecules, and that ribosomal RNA is the molecule of choice (Chapter 3). The procedure is simple in principle, though rather less so in practice. Pairs of sequences from different organisms are aligned, and the differences between them counted; their number is considered to be a measure of evolutionary distance. It bears repeating that such distances measure only the number of changes, not time elapsed, and that the rate of sequence change can vary between lineages and even within a single lineage. Pairwise differences between many organisms are combined to construct a phylogenetic tree, a hypothetical map of the pathway that generated the contemporary sequences (and, by implication, the organisms that house those sequences). One can even deduce the temporal order in which organisms diverged from their common ancestor, thus turning a phylogenetic tree (Fig. 3.2) into a statement about evolution over time (Fig. 8.2). Various algorithms have been devised for these purposes, each with its own virtues and defects, and with sequencing of genes and even whole genomes now routine, a veritable forest of phylogenetic trees is springing up. Not all the trees are congruent, but the windstorm of data has blown away the fog, revealing the bacterial world in a fresh and starkly molecular perspective (5). The most prominent feature in that landscape is the division of prokaryotes into two fundamental kinds. Archaea make up a new high-level taxon of organisms, profoundly different from the Eubacteria and of equivalent taxonomic rank. Microbiologists, at least, generally accept the designation of three domains, each of which is superimposed upon lower ranks that correspond to the traditional kingdoms (6).

And so we must rephrase the question that Stanier and van Niel thought they had answered fifty years ago: What manner of organisms are they that we designate Eubacteria and Archaea? This is not at all a simple question, for both domains contain a vast range of forms and lifestyles. Table 8.1 drawn from a burgeoning literature (7) summarizes the present state of knowledge, but is bound soon to be superseded by the torrent of information generated by whole-genome sequencing. Let me underscore here that Archaea and Eubacteria share the prokaryotic pattern of cellular organization, for that bland fact has turned out to be rather meatier than anyone realized just a few years ago. It goes beyond negative characteristics, such as the absence of intracellular organelles,

Table 8.1: Eubacteria and Archaea Reflected

Eubacteria	Common Features	Archaea
	Circular genome, no introns	
	Genes arrayed in operons	
	Basic machinery of DNA replication and recombination	
	Ribosomes 70S	
	Suite of ribosomal RNAs and proteins	
	Rigid helical flagella and rotary motor	
	Redox chains, cytochromes	
	Proton-translocating ATPases	
	Proton and sodium-coupled porters	
	Lithotrophic metabolism	
	Many catabolic and anabolic enzymes	
	Cross-linked cell walls	
	Division rings	
Eubacterial signature sequences in ribosomal RNA		Archaeal signature sequences in ribosomal RNA
Eubacterial transcription and translation (simple RNA polymerase, sigma factors particular promoters etc.)		Archaeal transcription and translation (multisubunit RNA polymerase, general transcription factors, particular promoters
DNA-binding proteins		Histones
Flagellin flagella		Glycoprotein flagella
F_1F_0-ATPase		A_1A_0-ATPase
Fatty acyl ester lipids		Isopranyl ether lipids
Peptidoglycan		Pseudopeptidoglycan
Bacteriochlorophyll		Bacteriorhodopsin, methanogenesis
FtsZ		FtsZ homolog

nuclear membrane and cytoskeleton. They share an array of specific and positive characters: circular genomes, 70S ribosomes, rigid flagella with basal rotors, chemiosmotic energy transduction with H^+ and Na^+ as coupling ions, cell walls of both the crosslinked and S-layer varieties and a surprising number of metabolic enzymes and pathways. In the archaeal genomes sequenced to date, about half the genes have homologs among the Eubacteria, and some Eubacteria harbor genes that resemble those of Archaea. Shared features may have been inherited from a common ancestor of the two domains, but there is good reason to believe that massive transfer of genes has contributed to their convergence (see below).

Differences become visible when one turns to the molecular elements, and particularly those concerned with the processing of genetic information. The digital sequences and signatures of ribosomal RNA which first defined the domains have no obvious functional significance, but serve as markers that track the "genetic core" of the organisms concerned. The patterns of DNA replication and of transcription and translation seen in Archaea are clearly different from those of Eubacteria, and the former have a distinctly "eukaryotic flavor." The presence of histones and proteasomes (specialized organelles for protein degradation) reinforce the impression that Archaea and Eukarya are sister groups, more closely related to one another than either is to the Eubacteria.

Looking down the list, one notes other differences that serve as taxonomic markers, but may also hold clues to physiological differences between the two kinds of prokaryotes (8). The crosslinked cell walls of Eubacteria and Archaea differ chemically: the former are always made of peptidoglycan, the latter of a similar but distinct polymer called pseudopeptidoglycan. Whether these chemical structures affect the physical parameters of the fabric in ways that are biologically relevant is presently unclear. The chemistry of membrane lipids is more suggestive. Both Eubacteria and Archaea employ phospholipids, but those of Eubacteria (and of eukaryotes) are made from the familiar fatty acyl esters depicted in every textbook of biochemistry, while Archaea make theirs of isopranyl ether lipids. These covalently-linked ether membranes are more resistant to hydrolysis and less likely to come apart at high temperatures. They may represent one of the adaptations that set the Archaea on an evolutionary trajectory of their own, but it is also possible that isopranyl membranes are a primitive character, inherited from the common ancestor of all prokaryotes, that was replaced in the other lineages.

Turning now to energy production, we note again the theme of mechanistic diversity emerging from a deep unity of organization. The many differences between the two domains are plainly adaptive, and seem to have arisen subsequent to their separation. Both Eubacteria and Archaea feature redox chains, ATP synthases and chemiosmotic energy transduction by proton currents, and both domains contain organisms that rely on inorganic reactions for energy, such as the oxidation of H_2S or of ferrous iron. But members of the two domains are likely to access energy sources in different ways. The Eubacteria seem to have taken full advantage of the availability of pre-formed organic substances; organotrophy, as illustrated by *E. coli*, is much less prominent among the Archaea. It is also the Eubacteria, and they alone, that invented the familiar mode of photosynthesis based on bacteriochlorophyll. One division of the photosynthetic bacteria then gave rise to the cyanobacteria, and with them to the oxygen-producing pattern of photosynthesis. The cyanobacteria went on to generate the bulk of the atmosphere's oxygen, and became the progenitors of eukaryotic chloroplasts. Archaea explored alternative metabolic universes. They are characteristically lithotrophs, earning their living by the oxidation of hydrogen with either sulfur compounds or CO_2 as electron acceptors. The latter pathway, in which CO_2 is reduced to methane, features coenzymes not encountered elsewhere; it is confined to the Archaea and specifically to a deep branch of that domain, the Euryarchaeota (Fig. 8.2). The same domain boasts another unique invention: the halobacteria carry out photosynthesis of a sort with the aid of bacteriorhodopsin, a light-harvesting proton pump totally unrelated to chlorophyll. The majority of modern Archaea, (although by no means all of them) inhabit environments that seem to us extreme: devoid of oxygen, strongly acidic or saline, deficient in organic nutrients and often extremely hot with temperatures close to and even above the boiling point of water. The typical archaeon is likely to be a lithotroph, an anaerobe, and a thermophile. The open question is whether these are adaptations to hostile niches in which the versatile Eubacteria do not thrive, or primitive characters inherited from an ancestral form. Archaebacteria were originally so named because they were expected to be relics of an early phase of biological evolution. A glance at Fig. 8.2 demonstrates that this cannot be true, but one can still make a case that Archaea hold clues to the kinds of environments in which prokaryotes first evolved.

The two domains are believed to have diverged two or even 3 billion years ago (see following section), and then diversified massively into novel habitats. This transformed all aspects of cellular operations, but not nearly to the same degree. The two modes of genetic information processing, one characteristic of Eubacteria and the other of

Archaea and Eukarya (Table 8.1), have remained distinctly different. It is not entirely clear what constrains divergence within the genetic core, but it probably has to do with a system made up of numerous components that must interact with high precision, and whose function has remained essentially unchanged. Metabolism, physiology and morphology were much more at liberty to vary and adapt to novel niches and changing circumstances. Norman Pace (7) lists several examples that illustrate that even within a given domain, phylogeny and physiology are loosely coupled. The gamma subgroup of the division Proteobacteria is represented in Fig. 8.2 by the purple bacteria. One of these, *Chromatium vinosum*, is a photosynthetic bacterium that uses H_2S as its external reductant. A much more familiar member of that same lineage is *E. coli*, everyone's favorite organotroph. And that branch also bears the symbiont of the tubeworm *Riftia*, a prominent resident of the oases that surround deep ocean vents; the symbionts oxidize H_2S and confer a lithotrophic lifestyle upon the animals that harbor them. (Animal mitochondria also stem from the Proteobacteria, in this instance the alpha subgroup). For a second example, a branch of the Gram positive bacteria bears clostridia, spore forming anaerobes notorious for causing gangrene. Within that same cluster one finds that giant bacterium *Epulopiscium fishelsonii*, the size of a large ciliate, whose offspring do not arise by cell division but develop in the mother cell's cytoplasm and eventually escape through a tear in her envelope. The Archaea, also, enjoy wide evolutionary latitude as documented by the far reaching differences between the genomes of two methanogens, *Methanococcus jannaschii* and *Methanobacterium thermoautotrophicum*.

The unconformity between phylogeny and physiology is surely telling us something important about constraint and innovation in cell evolution, but it is not clear at present just what it means. For the purpose of establishing a sound phylogenetic tree, the most reliable guidance comes from the genetic core, including the ribosomal RNAs. From this perspective, Archaea and Eubacteria represent sharply divergent lines. But it is simply not true that (as we believed a few years ago) Eubacteria and Archaea are as different from one another as each is from the Eukarya; aside from their genetic cores and membrane lipids it is hard to say what distinguishes the one stem from the other. Whether the creation of a separate domain for the Archaea is necessary, or even warranted, has become the subject of a sharp debate, whose outcome may turn on the nature of the progenitor that gave rise to all the modern prokaryotes.

The Ultimate Ancestor

Molecular science, for all its no-nonsense airs, asks one to swallow some real humdingers, and none bigger than the assertion that all extant organisms have descended from a unique population of cells in the distant past (in principle, from a single ancestral cell). This hypothetical organism is referred to as the last common ancestor (more technically, the cenancestor), and is represented on the universal tree of life by that first branch point where the line of descent that gave rise to the Eubacteria diverges from that which led to Archaea and Eukarya (Fig. 8.2).

The postulate of a single universal ancestor, its biblical overtones notwithstanding, rests on a solid foundation of fact. All organisms share a considerable number of basic molecular and organizational features that cannot be explained as a consequence of chemical necessity: ATP and pyridine nucleotide coenzymes, DNA and RNA, proteins constructed from a standard suite of amino acids, ribosomes, ion-translocating ATPases, lipid membranes and many more. The most compelling argument comes from the discovery that all extant organisms employ the same genetic code. In the absence of good chemical reasons why CUU should spell leucine while CCU spells proline, the only persuasive explanation is that the code as we know it was already a feature of the last common ancestor, and has been retained ever since because mutations that alter codon assignment are apt to be lethal. This cannot be strictly true because some variations have been found in codon usage by mitochondria and protists, but there are reasons to set these aside as special cases that do not fundamentally challenge the universality of the code.

To draw a more specific portrait of that ancestral organism, one must go out on a limb; the nature and placement of the limb are determined by the root of the universal tree. The current consensus (Fig. 8.2) puts the root (i.e., that first branch point) between the Eubacteria and the Archaea, making the latter a distant but specific sister group to the Eukarya. Rooting the universal tree is difficult because standard procedures, that rely on comparison of a set of related sequences with a distant "outgroup," are not applicable: there is no possible outgroup, and so the tree of ribosomal RNAs cannot be rooted. The first investigators to meet the challenge (9) took advantage of the rare cases of duplicated protein-coding genes that are represented in all three domains. For example, the α and β subunits of the proton-translocating ATPases (Fig. 5.4) are encoded by genes that duplicated very early, even prior to the common ancestor of all the ATPases; comparison of sequences from all three domains puts the Eubacteria on one side of the divide, Archaea and Eukarya on the other. Congruent rootings have been inferred from the genes for EF-Tu (a cofactor in ribosomal protein synthesis) and for aminoacyl-tRNA synthetases. We cannot here go into all the ifs and buts (10); however, it is necessary to note that rooting the tree requires the algorithms to be stretched to their limit, and that an increasing fraction of the trees derived from protein sequences flatly contradicts the conventional wisdom. The consensus rooting should, therefore, be taken as provisional and open to reconsideration.

We proceed with caution on the premise that the rooting is correct, and that therefore the common ancestor would have displayed attributes shared by contemporary Eubacteria and Archaea (plus, perhaps, some that are today seen only in one domain, having been lost from the other). The case has been argued in some detail by Doolittle (7), who concluded that the cenancestor must have been equipped with essentially modern DNA-based genes, and with the molecular apparatus required for the replication of genetic information and its regulated expression. Ribosomes, transfer RNAs, polymerases and repair enzymes were already standard parts. The hypothetical cenancestor probably featured a cross-linked cell wall, though its chemical structure is questionable. The wall, in turn, implies turgor pressure, morphogenesis by surface-stress, and cell division by septation (Chapter 6). In fact, we know that some Archaea contain homologs of the FtsZ protein, suggesting that a division ring may have been one of the ancestral features of the earliest prokaryotes.

Concerning ancestral patterns of metabolism and energy production, little can be said with confidence, but what clues there are point in a surprising direction. Most of the modern bacteria familiar to us live by the degradation of pre-formed organic matter, but this would have been scarce prior to the rise of higher plants. A far more abundant source of energy is sunlight, which once led Woese to suggest that the earliest prokaryotes may have made their living by photosynthesis. Contemporary Eubacteria feature several versions of chlorophyll-based photosynthesis, but these are absent from the Archaea, it is therefore unlikely that the common ancestors were phototrophs. That leaves the earth itself as the most likely provider of energy, carbon and other nutrients, all of geochemical origin (11). Perhaps the ancestral pattern of metabolism, like that of some contemporary prokaryotes of both domains, was chemolithotrophy: the oxidation of H_2 gas, H_2S or ferrous iron, with carbon monoxide, nitrous oxide or sulfur serving as electron acceptor. (Oxygen, the most common oxidant in the modern atmosphere, would have been virtually unavailable prior to the advent of cyanobacteria, which introduced oxygen-generating photosynthesis. Sulfate and nitrate must also have been scarce.) This speculation draws support from recent research on the molecular phylogeny of respiratory chains and ion-transport ATPases (11). One can make a good case for the proposition that electron transport chains, ATP synthase and chemiosmotic energy transduction were all part of that ancestral cell's physiology. The last common ancestor looks more and more like a proper bacterium that would not strike one as obviously primitive, and that must itself have been the product of a lengthy evolutionary history.

The argument is logical but the emerging portrait of the cenancestor does not ring true, if only because it offends one's intuitive expectation that primordial cells must have been primitive and markedly less sophisticated than their descendants. For this and other reasons, Woese has recently come to the conclusion that the very notion of the cenancestor as a discrete kind of organism is erroneous (12). He argues instead that all prokaryotes evolved from a population of protocells that were still at the "progenote" stage. Their genetic machinery would have been far less accurate than that of any contemporary organism, hampered by rudimentary mechanisms of translation and replication and beset with frequent errors. The metabolism and architecture of the progenotes would likewise have been underdeveloped; they may, for example, have lacked cell walls. Woese now envisages the cenancestor, not as an organism but as a miscellaneous community of protocells that frequently exchanged primal genes and evolved as a unit. Genetic complements were not fixed, but subject to continuous remodeling by mutation, rearrangement, and unregulated traffic in genes. Over time functional systems would have "crystallized" into successful configurations, and therefore become less receptive to the import of novelty; the first modules to emerge were probably those concerned with the processing of genetic information. In consequence, the fluid population differentiated into a small number of stable types, among which were the progenitors of the three domains. Woese's argument is rich in implications for the very nature of early evolution, and we shall revisit it in later chapters.

Now that we are well launched upon conjecture, it is tempting to peer around the next bend. Over the past decade, microbiologists have turned up a profusion of organisms that inhabit deep, dark and hot locations—submarine volcanic vents, long-isolated aquifers, even bare basalt rocks thousands of feet beneath the Columbia Plateau. Most of these organisms are chemolithotrophs, and they call into question our habitual perception of what makes normal habitats for prokaryotes. Some suspect that, contrary to conventional wisdom, it is not true that the bulk of the earth's biomass is in its forests, and that it is ultimately the sun that sustains life; we simply do not know how much life these subterranean habitats support (13). In any event, the new discoveries mesh with the observation that the deepest branches (i.e., the earliest divergences) of both the eubacterial and the archaeal domains consist chiefly of hyperthermophiles—strange bacteria that flourish at temperatures up to 113°, and under pressures of a hundred atmospheres and more. Could the last common ancestors have been hyperthermophiles? Indeed, should the origin of life be sought, not in Darwin's warm little pond but in some hellish cavern deep beneath the sea? The proposition has been vigorously defended and as fiercely rejected by those who see hyperthermophily as a remarkable adaptation, rather than an ancestral state (14); and I propose to leave it there, with a question mark.

By definition, the last common ancestor marks the first divergence that gave birth to the two prokaryotic domains. When did that epochal event take place? Ribosomal RNA sequences cannot tell for they evolve at varying rates; protein sequences make better clocks, but seldom reach as far back in time. All the same, by bold yet judicious extrapolation, Doolittle and his associates arrived at a tentative date for the separation of Eubacteria from Archaea between 3.1 and 3.8 billion years ago (15). That estimate meshes well with the fossil record of early microbial life which, albeit fragmentary, is now quite extensive. It consists of the macroscopic remains of mat-forming microbial communities called stromatolites, together with well-preserved assemblages of individual cells from the Archaean aeon, 2 billion years old and more (16). Recent explorations in northwest Australia brought to light an impressive collection of specimens (Fig. 8.3) dated to 3.5 billion years ago. Judging by their size and morphology, these most ancient cells known at present were all prokaryotes and remarkably similar in appearance to contemporary cyanobacteria, it is a reasonable guess that they grew photosynthetically and liberated oxygen as a byproduct. If we can take the geologists' word for the date of their rocky matrix, then the cellular ancestors of us all must have flourished well before 3.5 billion years ago, perhaps as early as the organisms that left their traces in Greenland rocks estimated to be 3.8 billion years old (16). What would we make of them if we could examine a specimen?

Another question to tickle the imagination is, What caused that first bifurcation? Koch (17) argues that divergence into distinct and permanent lineages must have been the result of critical inventions that

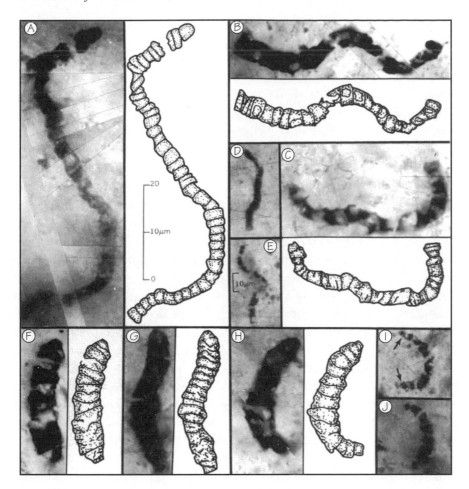

Fig. 8.3. Fossil filamentous prokaryotes from the apex chert of northwestern Australia. They are about 3.5 billion years old and appears to be cyanobacteria. Magnification for panels D, E, I, and J, see scale in panel E; for the others; see scale in panel A. From Schopf, 1993, with permission of the American Society for the Advancement of Science. Original photomicrograph courtesy of Dr. William Schopf.

gave their possessors a competitive advantage, or opened up new habitats. He suggests plausibly that Eubacteria may be the beneficiaries of high tech peptidoglycan walls, a product well suited to the conflicting demands of osmotic stability and safe enlargement; peptidoglycan walls are still a hallmark of the Eubacteria. As to the Archaea, Koch's surmise that methanogenesis gave them their edge is inconsistent with the observation that methanogens are found in only one of the known divisions of Archaea (Fig. 8.2), but perhaps not fatally so. A more serious defect of Koch's scenario is that it offers no hint as to why Archaea and

Eubacteria should have developed such very different mechanisms for processing genetic information.

Divergence is one side of the evolutionary coin, convergence the other. If 50 million years were sufficient for the ancestors of whales to return to the sea, shed their legs and sprout fins like those of fishes, what marvels of convergence may have taken place among prokaryotes in a billion years? The flagellar apparatus may be a case in point. Flagella of Eubacteria and Archaea are chemically quite different (18). It is hard to imagine them sharing a common ancestor, but not implausible that rigid helical rotating flagella may have arisen independently on two occasions. A deeper conundrum is presented by instances in which Eubacteria and Archaea feature closely related proteins, despite the evolutionary gulf that separates those two domains. The Eubacterium *Enterococcus hirae* produces both a proton-translocating ATPase of the eubacterial sort and a sodium-translocating ATPase whose affinities are with the Archaea (18); gene transfer from one domain to the other is the most likely explanation here. But what to do about the growing list of metabolic enzymes of clearly eubacterial kind found among the Archaea? Phylogenetic trees for glutamate dehydrogenase, glutamine synthase, heat-shock proteins and several others strongly suggest that, contrary to the paradigm of three separate domains (Fig. 8.2), the two kinds of prokaryotes should be grouped together; and that the archaebacteria emerged from a lineage within the Gram-positive eubacteria (19)! A few years ago, such discordant findings might have been set aside; but their number is growing as whole genomes tumble before the automated sequencing machines, and the challenge must be taken seriously. What is at stake here is nothing less than the reality of separate domains, and with it our entire conception of the prokaryotic world (20).

There is clearly something misleading about the crisp image of diverging lineages projected by the ribosomal RNA sequences. One possibility is that the conventional three-domain structure is erroneous. Archaebacteria may be a distinctive stem of the bacterial world, perhaps derived from the Gram-positive bacteria, and the features that set them apart no more (and no less) fundamental than those that distinguish the architecture of the Gram-positive envelope from that of the Gram-negative bacteria. Alternatively, we can attribute the presence of so many eubacterial genes in the genomes of Archaea to massive lateral gene transfer, involving a substantial fraction of the donor's gene complement. Here is where Carl Woese's vision of "genetic annealing" at the base of the universal tree of life (12) really comes to the rescue. The data emerging from the laboratories seem quite compatible with a population of progenotes engaged in the promiscuous exchange of genetic information. But the cooperative enterprise did not endure: as functional systems were refined, the barriers to the assimilation of imported genes grew ever higher. "Crystallization" began with the genetic core, while genes for the smaller metabolic modules could still find new homes. The genetic free-trade zone fragmented into protected enclaves, not abruptly but gradually on a time-scale of millions of years. If this is correct, then the peculiar distribution of genes in contemporary organisms is a relic of a communal genetic economy that passed away billions of years ago. Moreover, in the absence of discrete organisms and lineages, evolution in its earliest stages was quite unlike that with which we are familiar: its topology had the character of a net rather than a branching tree.

A Most Peculiar Alliance

The origin of the eukaryotic cell is arguably the most significant episode in the development of life on this planet, and surely the most baffling one. It is also not a single event, but a protracted process, whose roots reach deep into the early history of cellular life. There is no pertinent fossil record to speak of, so those who would plumb the origin of eukaryotes must rely on present-day molecules and "model" organisms. Fig. 8.4a illustrates the textbook consensus, sketchy and lightly held, that makes three claims. First, that Eukarya (like Eubacteria and Archaea) comprise a single lineage or clade: all eukaryotic organisms descend from a common ancestor. Second, that the eukaryotic lineage is nearly as ancient as the two prokaryotic ones, and is more closely allied to the Archaea than to the Eubacteria. And third, that advanced eukaryotic cells are chimeric in nature; they arose from the merger of early eukaryotic cells with certain Eubacteria, which became the precursors of organelles. The term Archezoa designates a kingdom level class of anaerobic protists that are plainly eukaryotic in structure but lack mitochondria and chloroplasts; these are believed to have diverged from the main eukaryotic stem prior to the acquisition of mitochondria. But now there is a growing sense that the ancestral eukaryotic cell was itself the fruit of unusual, even unique, events entailing the transfer of genes on a large scale across kingdom boundaries (Fig. 8.4b). The torrent of genomic data that is beginning to break over us may bring more certain knowledge, but in the meantime there is no choice but to let the imagination roam.

The first major breakthrough in the search for a eukaryotic genealogy can be credited to a single individual, Lynn Margulis, who maintained

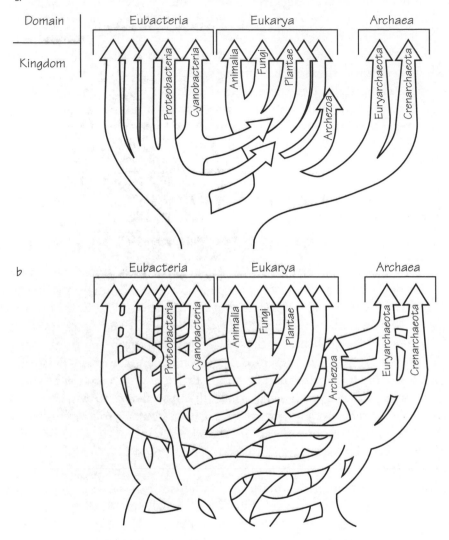

Fig. 8.4. Groping towards consensus on cell evolution. (a) The standard model, corresponding to the universal tree of Fig. 8.2. Three organismic stems arose from a common ancestor, giving rise to the Eubacteria, Archaea and Eukarya. Eubacteria, which had diverged at the base of the tree, re-entered the eukaryotic cell as endosymbionts, giving rise successively to mitochondria and then to plastids. Archezoa diverged from the eukaryotic stem prior to the acquisition of endosymbionts. (b) The evolutionary net. The early history of life is represented as an era of extensive gene exchange, out of which the three organismic stems crystallized. The status of the Archezoa is uncertain. After Doolittle, 1999, with permission of the American Society for the Advancement of Science.

that mitochondria and chloroplasts are derived from bacteria that had taken up residence in a primordial eukaryotic cytoplasm (endosymbiotic bacteria, in technical parlance). The proto-eukaryotic partner was envisaged as an anaerobic organism capable of engulfing bacterial prey; in some instances those bacteria, resisting digestion, established a more lasting relationship with their host. One set of bacteria contributed the capacity for respiration, an asset at a time when oxygen was accumulating in the atmosphere (Fig. 8.1). A subsequent acquisition of cyanobacteria brought in photosynthesis, and the successive consortia prospered thanks to their superior access to energy. The proposition that mitochondria and chloroplasts descended from endosymbiotic bacteria was not novel, having been mooted as early as the turn of the century, but it remained beer-parlor science until 1970, when Margulis (21) laid out the evidence in compelling detail and thus made cell evolution a respectable branch of scientific scholarship.

Resistance to the endosymbiont hypothesis persisted for another decade until ribosomal RNA sequencing sealed the matter (22). Mitochondrial rRNAs clearly belong to the eubacterial stem, specifically to the α-subgroup of the proteobacteria. Incidentally, that same ancestry had been inferred earlier by John and Whatley, who were impressed by the near-identity between the respiratory chain of animal mitochondria and that of a familiar aerobic Eubacterium, *Paracoccus denitrificans*. The origin of chloroplasts is placed just as firmly within the cyanobacterial lineage, confirming earlier proposals based on the mechanism of photosynthesis. Debate continues over how many independent episodes of symbiosis gave rise to today's organelles. Current opinion (23) favors the argument that each happened only once, making all mitochondria the descendants of a particular proteobacterium and all chloroplasts the offspring of a particular cyanobacterium. Bacterial endosymbionts are common enough among present-day eukaryotes, it is their conversion into organelles that makes the high barrier. Eukaryotic cells differ, of course, in their complement of erstwhile symbionts. Animal cells and most protists have mitochondria, while plant cells and photosynthetic protists have both mitochondria and chloroplasts. Many protists acquired chloroplasts secondarily, by assimilating another eukaryotic cell whole. Peroxisomes and hydrogenosomes are also thought to be derived from former endosymbionts, and some would assign such an origin to undulipodia and the cell nucleus (21). In a phrase coined by F. J. R. Taylor many years ago, the eukaryotic cell appears to be the product of *serial* endosymbiosis.

From prokaryotic symbiont to eukaryotic organelle is a tortuous road, entailing gains, losses and restructuring for both parties. Consider the mitochondrion. In order to serve as the cell's powerhouse, the symbiont must have acquired the transport system that exports newly made ATP from the mitochondrial matrix in exchange for ADP from the eukaryote's cytosol; such a device would have been worse than useless to a free-living microbe. The symbiont lost its cell wall, but not all of its envelope: mitochondria, like the Gram-negative bacteria from which they came, feature an outer membrane with proteinaceous pores, in addition to the highly convoluted inner membrane (equivalent to the plasma membrane) that bears the apparatus for oxidative phosphorylation. Most of the symbiont's genes ended up in the host's nucleus, which demanded the development of machinery to import proteins from the cytosol both into and across the two mitochondrial membranes. Mitochondria do retain a small genome that encodes some of the respiratory enzymes, plus the capacity for endogenous protein synthesis (with 70S ribosomes, like those of prokaryotes). The transformation of symbiont into organelle was surely a gradual and protracted process, that may have taken place *pari passu* with the diversification of the Eukarya. Contemporary mitochondria are not all alike, some retaining more of their ancestral genome than others. And in some cases mitochondria underwent still further transformation into specialized organelles, including the hydrogenosomes of anaerobic protists.

Endosymbiotic cyanobacteria likewise experienced both reduction and subsequent divergence on the way to chloroplast status. They retained thylakoids, the characteristic photosynthetic membranes of cyanobacteria, but acquired a clutch of modified chlorophylls and ancillary pigments, as well as new transport systems. True chloroplasts have lost their cell wall but not their outer lipid membrane. Chloroplast genomes are much larger than mitochondrial ones, but they, too, are no longer capable of independent existence. Their multiplication is controlled by the host cell, which retains the upper hand in any conflict between the nuclear and the organellar genomes. A curious aspect of all these transfigurations is that, while genes are subject to transfer and proteins come and go, membranes are commonly preserved; it is not at all clear to me why that should be so.

Thus far the ground is solid, but the footing turns mushy when one asks what manner of cell hosted these eubacterial symbionts; for this raises the question what really are the essential characteristics of a eukaryotic cell? The hypothesis as first formulated (21) assumed that the proto-eukaryote was capable of phagocytosis, implying the possession of a cytoskeleton, endomembranes and possibly a true nucleus. Protists with these features still survive: Cavalier-Smith (24) designated the kingdom Archezoa to accommodate anaerobic protozoa that are plainly eukaryotic in their general organization, but lack mitochondria and presumably never had any. Judging both by their ultrastructure and by molecular phylogeny, organisms such as metamonads (including the notorious parasite *Giardia lamblia*) are more closely related to the ancestral eukaryotic cell than are all other eukaryotes (Fig. 8.2). Recent molecular findings call the status of these organisms into question (see following section). All the same, the existence of such minimal protists highlights the key question about eukaryotic origins: Whence came cells endowed with a true nucleus, endomembranes, a cytoskeleton and undulipodia? The answer is unknown, leaving a huge lacuna in any account of cell evolution, but fostering a crop of stimulating conjectures.

In Margulis' view, the ancestral eukaryotic cell was itself the product of earlier mergers. The argument is spelled out in detail in her book *Symbiosis in Cell Evolution*, and also by Dyer and Obar (21), from whom I have borrowed the essentials of Fig. 8.5a. This illustrates one particular version of the serial endosymbiosis theory. The ancestral eukaryote is envisioned to be the fruit of an early fusion between an anaerobic, thermophilic, and wall-less prokaryote, (resembling, perhaps, the contemporary archaeon *Thermoplasma*) and a motile eubacterium such as a spirochaete. The product of this union, a primitive anaerobic flagellate, inherited from its progenitors the rudiments of what later became the hallmarks of true eukaryotic cells: the genetic core, histones and actin precursors from the archaeon, metabolism and propulsion from the eubacterium. Transformation of this primordial partnership into a recognizable eukaryote is implied but not spelled out. Subsequent fusion with an aerobic eubacterium (exemplified in the diagram by *Paracoccus*) gave rise to mitochondria and generated cells ancestral to those of animals and fungi. The acquisition of cyanobacteria in yet another round of fusion launched the photosynthetic protists and plants. Margulis still insists that the classification of organisms into kingdoms must grow out of this prehistory of genome mergers, and that the traditional scheme of five kingdoms meets the requirement.

Molecular phylogenists, who draw their opinions from the bedrock of gene sequences, view the matter somewhat differently but still in a glass, darkly. Recent authors (25) rightly underscore the compelling evidence for the existence of three fundamental lineages and assign particular significance to the apparent family links between Eukarya and Archaea. They argue convincingly that molecular phylogeny renders untenable the traditional five kingdoms, one prokaryotic and four

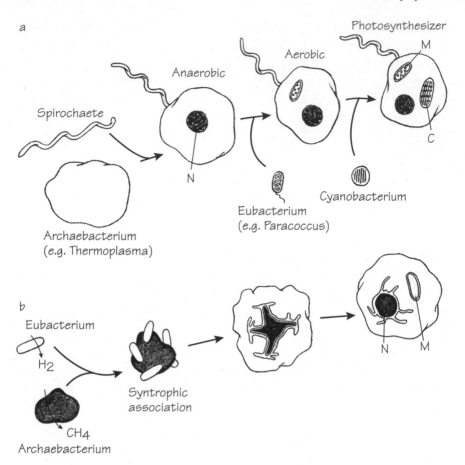

Fig. 8.5. The symbiotic origin of eukaryotic cells: two hypotheses: (a) Genesis of the basic eukaryotic types by serial endosymbiosis, as envisaged by Margulis and by Dyer and Obar. (b) A syntrophic model. The union of a methanogenic archaeon with a hydrogen-producing eubacterium leads to cytoplasmic fusion, membrane differentiation and elaboration of both nucleus and mitochondrion. Inspired by Martin and Müller, 1998, and Moreira and Lopez-Garcia, 1998.

eukaryotic, all roughly equivalent in rank and depth. It implicitly contradicts the proposition that ancestral eukaryotic cells arose from the fusion of prokaryotic cells akin to contemporary ones, since the origin of eukaryotic cells lies much further back in time than today's bacteria. The notion that undulipodia and such cytoskeletal elements as centrioles and spindle fibers should be traced back to a motile eubacterial symbiont (21), for which no evidence has ever been produced, is being superseded by the discovery that tubulin is homologous to the bacterial cell-division protein FtsZ (26; we still await solid evidence regarding the genealogy of actin). But far from resolving the issue of eukaryotic origins, molecular phylogeny has (at least for the present) deepened the mystery. The recent proliferation of protein-sequence trees suggests that the eukaryotic cell, rather than being a straightforward offspring of the Archaea, mingles genetic-information processing in the archaeal manner with cytoplasmic enzymes of eubacterial origin; and that reopens Pandora's box.

The better part of valor may be to sit tight and await the tide of new data, but only dullards are proof against the temptations of myth-making. Endosymbiosis, serial or otherwise, necessarily reemerges as the god in the machine (27). In perhaps the most radical of recent proposals, Mitchell Sogin traces the genetic apparatus of the eukaryotic cell to an organism of the archaeal lineage that fused with a surviving offshoot of Woese's hypothetical progenote. The latter, still at an RNA-based stage of genetic evolution, would be the source of RNA-splicing mechanisms and also of that puzzling cytoskeleton. Less venturesome authors would bring Eubacteria into the consortium, not as a motility symbiosis but as a souce of cytoplasmic enzymes. Recent indications that Archezoa contain genes of eubacterial provenance certainly encourages speculation along this line. And in quite a fresh offering, several investigators (27) have rethought the endosymbiotic scenario from the ground up. They suggest an association of two prokaryotes, an anaerobic archaeon that required hydrogen (a methanogen, perhaps) and a eubacterium that produced hydrogen under anaerobic conditions but was capable of respiration. What gave this partnership an initial selective advantage would be, not respiration but the tightly coupled transfer of hydrogen from the eubacterium to the archaeal partner. However, as photosynthetic bacteria generated increasing levels of atmospheric oxygen, the eubacterial partner was well positioned to evolve into the mitochondrion. This proposal represents a significant alternative to serial endosymbiosis: instead of a protoeukaryotic host that acquired bacterial symbionts (Fig. 8.5a), we now entertain the notion that mitochondria originated in tandem with the eukaryotic order itself (Fig. 8.5b). Note, however, that this stimulating notion leaves the status of chloroplasts unchanged (still of symbiotic origin), and sheds little light on the provenance of nuclei, endomembranes and the cytoskeleton.

There is a fine air of whimsy about those imaginative tales, with overtones of Rudyard Kipling ("And this, O Best Beloved, is why . . ."). They also step insouciantly around patches of quicksand, such as what brought about early cellular fusions that are not permitted to contemporary prokaryotes, why some genes were discarded and others

preserved, and how a consortium of prokaryotes acquired the architectural and functional complexity of even the simplest eukaryotic cell.
Perhaps what is wrong is that the tales are not imaginative enough! But
the speculative ferment is healthy, for it grows out of new information
and fresh approaches. What matters here is not the dubious particulars
of parentage, but the mounting indications that repeated mergers of
cells and their genomes lie at the base of the eukaryotic stem and remain
a feature of the entire lineage. Symbiotic fusion of organisms can in
principle explain the transfer, not merely of individual genes but of
articulated functions such as motility or phagocytosis; and it may hold
unforeseen clues to the origin of integrative patterns in general. And
these fusions and mergers, whose nature is still being worked out, lend
all of early cell evolution the character of a net, quite unlike the branching bush of more recent evolutionary times.

THE RISE OF THE EUKARYOTES

There are said to be more than 200,000 known species of unicellular
protists, and that is probably an underestimate. It is instructive to leaf
through Margulis and Schwartz's atlas of living phyla or the *Handbook
of Protoctista* (28) just to savor the variety of forms, habitats and live-
lihoods explored by the single-celled eukaryotes. Fig. 7.1 will remind
the reader of some relatively familiar creatures out of that vast bestiary:
amoebas, ciliates, phototrophic algae and the oömycetes which grow like
fungi but belong among the protists. The lower eukaryotes are apt to
be neglected, even by biologists. Yet among them are to be found, not
only many pathogens of animals and plants, but also the progenitors of
all the higher forms of life; and they display endless variations on the
principles of eukaryotic organization.

The classification and phylogeny of the protists, "nature's most out-
landish oddities" (as Lewis Thomas put it in his preface to *The Handbook of Protoctista*), has been beset with disputation ever since Haeckel
first created his kingdom Protista more than a century ago. What is
new is that, thanks to the advent of molecular phylogeny centered on
ribosomal RNA sequences, there is growing agreement on an objective
historical framework upon which biologists of a more organismal bent
can hang their own findings. Fig. 8.6 again depicts the universal phylogenetic tree, but with the emphasis on the eukaryotic stem. Its subsidiary branches display the order of bifurcations that gave rise to the
eukaryotic world as we know it, and provide a measure of the evolutionary distance between organisms as judged by their ribosomal RNA
sequences. To be sure, such trees are but skeletons of the green tree of

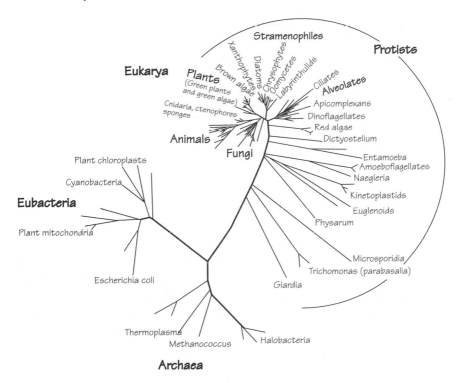

Fig. 8.6. Phylogeny of the Eukarya, in the context of the universal tree. The evolutionary distance between any pair of organisms is measured by the length of the lines that connect them. Note the density of the crown and the shallowness of the brushes that identify animals, plants and fungi in comparison to the earlier protistan lineages. Tree after Sogin, 1994, and Pace, 1997.

life, for they track the genealogy of a particular set of genes rather than of organisms, but they bring us as close to the phylogenetic truth as we are likely to come.

There is fair agreement on the order of most of the branches and their length, but not on the best way to subdivide the huge domain of the Eukarya into coherent and manageable taxons. Molecular phylogenists argue that, if animals, plants, and fungi are to retain their status as separate kingdoms (surely a practical necessity), then many of the clusters and deep branches within the protists demand equal status. Mitchell Sogin (from whose extensive studies Fig. 8.6 was drawn; 29) and several others have designated two such kingdoms, the stramenophiles (chiefly various algae and relatives that have lost their plastids) and the alveolates (ciliates, dinoflagellates, etc.). Cavalier-Smith would place the former in kingdom Chromista, the latter into Protozoa (29). Margulis, as mentioned above, argues for an altogether different basis for taxonomy and defends placing all eukaryotes other than animals, plants, and fungi into a multifarious kingdom, Protoctista, with some forty phyla. I intend to venture no deeper into this minefield, both for lack of expertise and because the genealogy supplied by ribosomal RNA sequences is sufficient for the purposes of this chapter.

It takes but a touch of imagination to read the phylogenetic tree of the Eukarya (unlike those of the two prokaryotic domains) as a historical record, displaying broad trends: diversification, to be sure, but also mounting size, independence and complexity. Note, however, that much of the organizational complexity is already present at the very base of the tree, in that ancestral eukaryotic cell of whose nature and origin we know so little. The profusion that came after is built like a fugue upon the deep theme of eukaryotic order.

From this point of view, it is the lowest and most ancient branches that are the most intriguing, for lineages that diverged prior to the acquisition of endosymbionts should hold clues to what the earliest eukaryotes were like. As mentioned above, Cavalier-Smith (24) designated a separate kingdom Archezoa to house such archaic-looking protists as the metamonads, parabasalians, archamoebae and microsporidians (Fig. 8.7). Obscure even by protistan standards, all the deep branches include members that are parasitic upon animals. Travelers and hikers are sure to learn about the metamonad *Giardia lamblia*, the causative agent of a most distressing gut infection in humans; and it is this interface with our own persons, rather than its ancient lineage, that has drawn the attention of researchers. Setting aside clear adaptations to a parasitic lifestyle, we note that *Giardia* contains no mitochondria, peroxisomes or hydrogenosomes, all thought to be derived from endosymbionts. It does feature undulipodia, basal bodies, a true nucleus enclosed in a nuclear membrane with pores, mitosis and a simple microtubule-based cytoskeleton. Tubulin, actin, myosin, centrin and other familiar cytoskeleton proteins have been reported. Recent studies show that, contrary to earlier belief, a well-developed system of endomembranes is present, complete with equivalents of the Golgi apparatus and lysosomes of higher eukaryotes (30). All this confirms the general opinion that endomembranes and vesicle trafficking evolved in the ancestral eukaryotic cell, together with the membrane-bound nucleus. *Giardia* has 70S ribosomes (like those of prokaryotes, and unlike the 80S ribosomes of higher eukaryotes), and also metabolic enzymes of the bacterial kind. It makes quite a plausible candidate for a minimal eukaryotic cell.

It is a little jarring to learn that a human parasite may be the direct

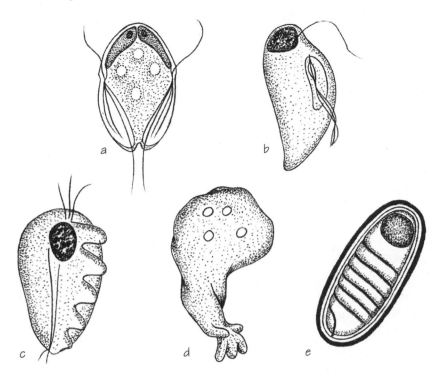

Fig. 8.7. A gallery of provocative protists. (a,b); Possible descendants of ancestral eukaryotes: the metamonads *Trepomonas agilis*, a free-living relative of *Giardia*, and *Retortomonas*. (c,d,e): Recently demoted: a parabasalian, *Trichomonas;* the antique-looking but fairly modern archamoeba, *Pelomyxa palustris*; and the microsporidian *Nosema*, a badly misplaced fungus. After Keeling, 1998, with permission of John Wiley and Sons.

descendant of an ancestral kind of eukaryote that flourished 2 billion years ago, but it probably should not bother us. *Giardia* does have free-living relatives, such as *Trepomonas agilis* (30; Fig. 8.7a), and parasitism is one of the alternative nutritional options open to organisms incapable of photosynthesis. A more serious question is whether the metamonads intrinsically lack mitochondria, or have lost them secondarily in the course of evolution. Recent research indicates that metamonads carry genes that are usually associated with mitochondria (31), and that parabasalians (Fig. 8.7c) contain hydrogenosomes that are derived from mitochondria. *Microsporidia* (Fig. 8.7e) appear to be misplaced altogether, being relatives of the fungi. If all these turn out to have once harbored endosymbionts, we may be forced to conclude that we know of no eukaryotes that inherently lack mitochondria! It need not follow that no such cells existed in the remote past; nevertheless, we must seriously entertain the possibility that mitochondria hark nearly as far back as the true nucleus does, and that bacterial symbioses lie at the very root of the eukaryotic mode of organization. Schemes such as that sketched in Fig. 8.5b may not be as far-fetched as they appear.

Protistan lineages in the middle reaches of the eukaryotic tree (Fig. 8.6) commonly contain mitochondria but not chloroplasts, they are aerobes, but can often thrive in oxygen-poor habitats. Their predominant nutritional mode is phagocytosis, but many can also take up dissolved organic matter. The glaring exceptions are the euglenoids, with photosynthetic members. It seems likely that these are basically protozoa that incorporated chloroplasts from a primitive plant, and thus acquired photosynthesis secondarily and quite late in their history (29).

Most of eukaryotic diversity is lodged in the densely branched "crown" of the eukaryotic tree, where we also encounter many of the most successful lineages (32). These include the alveolates (ciliates, dinoflagellates and the largely parasitic apicomplexans); the stramenophiles (a great concourse of protists and unicellular algae, most of them phototrophs and others, such as oömycetes, that lost their plastids secondarily), and of course the green plants, animals and fungi. Multicellularity emerged independently on several occasions, initially with the cellular slime molds and the red algae. Note that the great majority of photosynthetic organisms are found in the crown. Whether all extant chloroplasts derive from a single unique event or from several independent ones is still under discussion. However, even if all chloroplasts share a single common ancestor, it is clear that many protists acquired them independently by incorporating a phototrophic eukaryotic symbiont in its entirety. Remnants of the latter's nucleus sometimes survive, and (rather curiously) so do its membranes. The number of membranes that enclose contemporary chloroplasts is thought to betray the organelle's ancestry (32).

It is very tempting to read into the diagram (Fig. 8.6) a tale of cause and effect, perhaps along the following lines. Early anaerobic eukaryotes, restricted to fermentable substrates, were low in both abundance and diversity despite their ability to catch prey and engulf it. Their prospects improved with the acquisition of endosymbionts: in a world increasingly rich in oxygen thanks to the activities of cyanobacteria, mitochondria made energy freely available. And then, following the later acquisition of chloroplasts, energy ceased to restrict proliferation and the great radiation of eukaryotic life followed. There is likely to be some truth in this story, but it must never be construed to mean that early eukaryotes acquired endosymbionts *because* that put their energy budget in the black. In fact, we do not know what were the benefits that accrued to the host cells that first tolerated intracellular guests; perhaps we should consider endosymbionts as a kind of "pre-adaptation," whose full significance appeared only after prolonged cohabitation and mutual accommodation.

History without dates lacks authority, and every effort is being made to remedy the deficiency, but the deeper one peers into the past the fewer signposts stand forth (Fig. 8.1). One set of dates has been drawn from molecular phylogeny itself. As mentioned earlier, Doolittle and his associates (15) constructed a deep phylogenetic tree by pooling more than three hundred amino acid sequences from 64 different enzymes, and calibrated it by reference to the fossil record of animals. Extrapolation from that solid base into the mist indicates that plants, animals, and fungi last shared a common ancestor about one billion years ago. Eukaryotes diverged from ancestral archaebacteria around 2.3 billion years ago, and the last common ancestor of all cellular life goes back more than three billion years. This dating of the eukaryotic tree meshes reasonably well with the independent set of dates supplied by the fossil record.

According to the admirable surveys by Knoll (32), the "big bang" radiation of eukaryotic life took place a billion years ago or a little before. But fossils generally taken to be of eukaryotic nature are found in rocks very much older than that. Steranes, chemical fossils produced by the degradation of sterols (which are characteristic of eukaryotic cells but not of prokaryotes), are found in strata as old as 2.7 billion years. Large objects called architarchs, possibly resistant cysts produced by photosynthetic protists, go back 1.8 or 1.9 billion years. And a macroscopic fossil alga called *Grypania*, thought to resemble the contemporary *Acetabularia* (Fig. 7.1f), has been found in rocks dated to 2.1 billion years. This would put the acquisition of mitochondria well before 2 billion years ago, of chloroplasts not much later, and by 1.7 billion years ago eukaryotes were a significant component of the biological community. Why, then, would the radiation of eukaryotes be delayed by another half billion years? Something was still lacking, and Knoll suggests that this may have been effective sexual reproduction.

It is altogether remarkable (as Mark Twain observed on the Mississippi) what wholesale returns of conjecture science extracts from its trifling investment of fact. And we are far from done with the ancient mysteries, for even a full-leafed phylogenetic tree will tell us little about the origins of physiological processes and adaptive functions that must be the substance of evolutionary history. Some of that convoluted story can be read in the ultrastructure and life cycles of protists and fungi, for instance the many versions of mitosis or the variety of symbioses that knit the living world into an integrated economic community. But its kernel seems as baffling as ever: how a prokaryotic cell, or a consortium of such cells, turned into a radically different entity that we now call a eukaryote. More or less plausible accounts of how it may have happened have been put forward (33). Cavalier-Smith has definite ideas about the origin of endomembranes, a true nucleus and the cytoskeleton; de Duve presents a rather different version. And Maynard Smith and Szathmáry, in *The Major Transitions in Evolution*, go on to reflect on the origins of mitosis, sex, and societies. There is nothing whatever wrong with disciplined speculation—how else would we know what to look for in the ever-growing heap of facts and factoids? But it does warrant the "amiable cynicism" of the Italian maxim quoted by Roger Stanier (17) in one of the first modern essays on cell evolution: *Se non è vero, è ben trovato.* (It may not be true, but it's well contrived.) And one cannot help suspecting that we are approaching a limit to what can be known, set not by technology but by the nature of this inquiry into the inconceivably remote past.

9

By Descent with Modification

"All advances of scientific understanding, at every level, begin with a speculative adventure, an imaginative preconception of what might be true—a preconception that always and necessarily goes a little way (sometimes a long way) beyond anything which we have logical or factual authority to believe in . . . Scientific reasoning is therefore . . . a dialogue between two voices, the one imaginative and the other critical."

Sir Peter Medawar (1)

THE DARWINIAN OUTLOOK
ENLARGING THE ENVELOPE
THE GENERATION OF NOVELTY
A SINGULARLY SUCCESSFUL DESIGN
A DIRECTION TO EVOLUTION?

Everything that exists in the universe is the fruit of chance and necessity. Thus spake Democritus in the fifth century B.C.E. and Jacques Monod chose those key words for the title of his celebrated (and often infuriating) book on life in a universe without God. The phrase neatly encapsulates the central principle of evolution as Darwinians see it, adaptation by the interplay of random variation and natural selection, but it is evocative enough to admit of more than one reading.

Evolutionary history, like any other kind of history, must be practiced on two levels. The first requirement is to establish the phylogeny, the lines of descent that map out the relationships among organisms. That goal is presently the focus of intense research, whose status was outlined in the preceding chapter and in Figures 8.2 and 8.6. The second, and harder, task is to search out the causes that lie behind the universal tree of life. Some readers may protest that there are none, beyond that endless chain of "breedings and weedings" (a phrase attributed to the physicist Steven Weinberg). But this does not ring true, no more than the flippant dismissal of human history as one damn thing after another. There may be no *laws* of either kind of history, but there clearly are trends and regularities whose causes demand attention. Is every organismic feature to be understood as the product of adaptation by natural selection? Is the pattern of nature wholly accounted for by gene mutations generating a pool of variants for sifting, or do additional mechanisms contribute to the creativity of evolution? Why, indeed, are there so many kinds of organisms large and small, and why do they cluster into discrete species? The eukaryotic stem, and to a lesser degree the prokaryotic ones, display not only diversification and adaptation but a continual increase in organismic complexity; does this mean that evolution is in some sense progressive or directional? I shall reserve the origin of life for the final chapter of this book; there is plenty of food for thought on the road from the universal cellular ancestor to life as we know it.

Any reservations about the mechanisms of evolution are apt to be seized upon by enemies of the principle itself, and make Darwin's children anxious. Let me, therefore, state unambiguously that I, like the vast majority of contemporary scientists, see the living world as wholly the product of natural causes, including a continuous chain of descent with modification from a population of primordial cells. There is no evidence to indicate that it was shaped by the mind and will of an external creator. In making this affirmation, I am not just tipping my hat to scientific orthodoxy. For millennia, philosophers and plain folks alike have taken it for granted that the order we see in the world is rooted in some kind of transcendent, cosmic mind. Ever since Darwin, and only since Darwin, has an alternative viewpoint been available. The theory of evolution represents the most profound and most radical change in the way we look at the world since late antiquity, and its implications touch every facet of the life of reason (2).

That said, there remains a battery of open questions about the forces and events that shaped the history of life. The purpose of this chapter is to examine what we can learn from the early stages of that history about the process of evolution itself.

THE DARWINIAN OUTLOOK

In 1982, at the centenary of Darwin's death, Stephen Jay Gould published a trenchant article that managed both to celebrate the theory of evolution and to call for substantial revision of its central tenets (3). It is not, of course, the fact of biological evolution that Gould challenged, nor even the crucial contribution that natural selection makes to shaping the living world. The issue was whether "Darwinism," as represented by the body of thought referred to as the modern evolutionary synthesis, is sufficient to account for the pattern of nature. Gould argued that the framework of the day was too narrow and required expansion, particularly in its application to the origin of new species and the higher categories. In order to appreciate what is at stake here, let us begin with a quick excursion through the history of evolutionary thought. It is well warranted, for to anyone who accepts that evolution is the process that created the living world, there can be no more fundamental question than how evolution itself comes about.

Gould's article fired one salvo in a debate that had been ongoing for some years, and continues today, over just what the theory of evolution stands for. Briefly, evolution as a fact of biological history was generally accepted in Darwin's lifetime, but his contention that natural selection is the chief agent of evolutionary change (chief, not sole) only came to prevail long after his death. General acceptance grew out of the clarification of the nature of heredity and variation. Darwin and his contemporaries knew nothing of the mechanism of heredity; they thought of it broadly as all that accounts for the stable and repeatable nature of reproduction. Once it was accepted that heredity is effected by the transmission of particles called genes (after W. I. Johannsen coined the term in 1909), and that mutations in those genes represent the ultimate source of variation within populations, evolution by natural selection could be placed upon a quantitative mechanistic foundation. The modern synthesis refers to that fusion of Darwin's own ideas, drawn from the English tradition of natural history, with the novel science of population genetics. Consensus was achieved around the middle of the twentieth century by a galaxy of eminent scholars including Theodosius Dobzhansky, Ronald Fisher, Julian Huxley, Ernst Mayr, George Gaylord Simpson and Sewall Wright (3).

In the course of the modern synthesis, the very meaning of the term "evolution" underwent a substantial transformation. Gilbert, Opitz and Raff (4) summarize what happened. "In 1937 Morgan's student, Theodosius Dobzhansky, . . . took the bold step of redefining evolution as changes in gene frequency. Instead of being a phenotypic science analyzing changes in fossil morphology, embryonic structure, or the alterations that make a structure adaptive in a particular environment, evolution became the epiphenomenon of the genetics of populations." In a nutshell, evolution results from the interplay of two processes, variation at the level of the genes and selection among organisms for survival and reproductive success. In this indirect manner, natural selection promotes the spread of genes (most commonly of variants of a particular gene, called alleles) that enhance the adaptation of organisms to their environment, and inhibits the propagation of less efficacious genes. Mutations are the raw material upon which selection operates; they occur "at random," in the sense that mutations happen without regard to their prospective usefulness. In each cycle of reproduction (and occasionally at other times), genes undergo exchange and recombination. And in every generation, those organisms, which by virtue of their genetic patrimony are better adapted, are more likely to survive and to leave more offspring. That is what students of evolution mean by the term "fitness"; it has nothing to do with the ability to run the marathon. The interplay between chance and necessity builds better adaptations one small step at a time, with natural selection (5) as the chief directing force and arguably the sole such force. The very same processes that account for the spread of protective dusky pigmentation among woodland moths in grimy industrial England also explain the transformation of fishes into amphibians, or the origin of the eukaryotic cytoskeleton. Natural selection emerges as the preeminent creative force to which we owe all the marvels of biology.

Evolutionary science has been fortunate in its expositors: the writings of Richard Dawkins and Daniel Dennett (6) surely rank among the most instructive and most accessible contributions to the biological literature, guaranteed to raise readers' awareness and often their hackles too. What these books profess is a hardened version of the modern synthesis, in which the recurrent theme is the replication, variation and selection of *genes*, while organisms take the back seat. Genes (for that matter, strings of DNA generally) have the supreme virtue that they are copied exactly; unlike organisms, genes replicate (the difference between the replication of genes and the reproduction of organisms is familiar to any parent). It is the endless chain of DNA replications that links every organism to its forebears, and through them to the primordial cell that first made use of DNA as a genetic archive. Occasional errors of replication supply variation, and represent potential branch points in that river of digital information. From the genocentric perspective one can argue that the beneficiaries of the evolutionary game are, not the fitter organisms but the more successful genes. Bodies, with all their beautifully crafted adaptations, are but the instruments through which genes maximize their chances in the competition for reproductive success.

Richard Dawkins in particular is celebrated (and in some quarters, reviled) for his potent metaphor of the "selfish gene", that accentuates the idea that genes compete with one another and that bodies with all their adaptations exist for the good of their genes and not the other way around. Organisms do not, of course, vanish altogether. They serve as the vehicles in which genes are collected and exposed to the hammer of natural selection. It is important to remember that genes as such are invisible to selection, which notes only their manifestations as expressed in a phenotype. But selection *of* organisms often has the effect of selecting *for* a particular gene (7). And there are more than a few observations, particularly in zoology, that are very difficult to explain on any basis other than competition among genes. When I first encountered the startling proposition that evolution results from the struggle among genes rather than organisms, it struck me as both distasteful and wrongheaded. Two decades and many printed pages later, putting genes ahead of organisms no longer seems to threaten everything that I hold to be true. But it still makes for a narrow and one-sided view of the phenomenon of life.

Samuel Butler once famously described a hen as the egg's way of making another egg; and when one has become accustomed to the shift of viewpoint the quip is instructive as well as amusing. Physicists perceived the world afresh when they came to terms with the idea that light has the properties of particles as well as waves. In somewhat the same way, one can see organisms that enshrine their traits in the form of replicable genes, and genes that express selectable virtues in the phenotype of an organism, as complementary aspects of living systems. Which viewpoint is the more informative depends on the question at hand. For the purpose of thinking quantitatively about evolution, it is often convenient to treat the organism as a "black box," and to score genes as being more successful or less successful in the struggle for immortality, the nature of the lottery does not depend on the manner in which the genes interact to make the box. But those genes can only reproduce themselves as part of an organism, and this is the point of contact with the biology we experience every day. When one asks about adaptations, and how they arise in the course of growth and development, genes make too narrow a focus, and an integrative perspective is called for. Genes still have a role to play since they encode the proteins that underlie traits, but they no longer hog the spotlight. On the levels of physiology and development, gene products are so closely intertwined that there is usually no simple mapping between genes and traits; and whatever selfish proclivities genes may have are usually constrained for the good of the collective. Could organisms likewise constrain what is possible in evolution? Quite likely, and such a conclusion would not in principle conflict with the modern synthesis, whose commodious framework leaves ample room for exceptions, nuances and a diversity of viewpoints.

If there is not outright conflict between gene-centered evolution and organism-based physiology, it must be said that there is between them no consilience (8) either: one does not grow naturally out of the other. So long as all questions about the origin of organisms and their transformation over time must be answered by reference to changes in gene frequencies, there is good reason to suspect that something important has been left out of the reckoning. Some of the numerous attempts to bridge that gap will occupy us for all of the following section, beginning with Gould's gauntlet of 1982.

ENLARGING THE ENVELOPE

What Gould challenged in his 1982 article, and in others before and after (3), is not the validity of the modern synthesis but its sufficiency. As Gould sees it, the synthesis rests upon twin pillars. One is Adaptationism, the presumption that all (or at least, the great majority) of biological features represent adaptations selected for their utility. The other is Gradualism, the belief that all evolutionary change must be gradual since it results from the accumulation of small changes sifted by natural selection. Gould does not deny the strength of these two foundations, but argues that additional factors must contribute to the shaping of our world. "The modern synthesis is incomplete, not incorrect."

Does every one of an organism's traits enhance its fitness by offering a selective advantage? On the contrary, Gould insists that many features were not adaptive in their origin but arose as byproducts or incidental consequences of some other change that was favored by natural selection. A neat instance is the spiral space left open when a snail is shaped by winding a tube around an axis. Some snails make this space serve as a protected brood-chamber for their larvae, others make no use of it at all, but in either case the space exists for reasons geometrical, not for any selective advantage that it confers upon its possessor. Gould refers to such features as "spandrels", an architectural term that designates the curved triangular surfaces that enclose the space created by setting a round dome upon a square of arches. Spandrels are often sumptuously decorated, as are those of the Cathedral of San Marco in Venice, which inspired the term; but the reason that spandrels exist is structural, not decorative. Spandrels, then, are shorthand for the major category of biological features that do not arise as adaptations. That does not preclude their being subsequently co-opted or "exapted" for some useful purpose, as in those snails that make a virtue of architectural necessity.

The premise that evolutionary change is necessarily gradual also comes into question. Gould and Eldredge (3) created quite a stir when, nearly thirty years ago, they pointed out that the pattern of evolutionary change displayed in the fossil record is quite unlike the steady, slow transformation of populations that Darwin seems to have had in mind. Their term "punctuated equilibrium" identifies two unexpected features of the pattern. First, species appear in the record abruptly (abruptly, that is, on the geologist's timescale, where the record of a hundred thousand years may be compressed into a bedding plane just a few millimeters thick). Second, the organisms that make up a species commonly remain virtually unchanged for millions of years before going extinct. The reality of this pattern has been extensively confirmed for organisms as diverse as ancient horses and the microscopic protists called foraminifers. The authors, therefore, recast evolution as a tale about the differential success of species. "All substantial evolutionary change must be reconceived as higher-level sorting based on differential success of certain kinds of stable species, rather than as progressive transformation within lineages" (3). Gould and Eldredge would draw a line between two kinds of evolutionary change. One is microevolution, the spread of beneficial gene alleles within a species (for example, the mutant hemoglobins of humans that confer resistance to malaria, or the butterfly pigments that ward off predators). Macroevolution, by contrast, designates the large transitions that involve major innovations (feathers, or myelin) and underlie the divergence of the higher taxons. The mechanisms of macroevolution are not well understood, but appear to differ in kind from those that tailor genes one step at a time, and include some form of selection among species.

Gould's arguments, which I have sketched here in an abbreviated fashion that does less than justice to either their subtlety or their style, have stirred up a tempest among adherents of the hard-edged ("fundamentalist") version of Darwin's church. There is no time here for a tour of the battlefields, but I must confess to taking malicious pleasure in Chapter 10 of Dennett's *Darwin's Dangerous Idea* (clever and not entirely unjustified, but also unfair and in places cheerfully offensive), in Gould's measured formal rejoinder, and in an uninhibited exchange of brickbats in the *New York Review of Books* (9). Evolutionists, beginning with Darwin himself, have always recognized that both the mode and the tempo of change vary over time. Gould and Eldredge have, therefore, been accused of putting up straw men for easy demolition, and of reformulating their claims whenever they encounter criticism; and there is a grain of truth in these charges. But there is nothing evasive in the challenge to orthodoxy: "I think I can see what is breaking down in evolutionary theory—the strict construction of the modern synthesis with its belief in pervasive adaptation, gradualism, and extrapolation by smooth continuity from causes of change in local populations to major trends and transitions in the history of life" (3).

There is a theological fervor to this dispute (conducted among protagonists who, after all, are basically on the same side) that suggests that what is at stake here goes beyond mechanisms and even personalities. The modern synthesis is reductionist in the sense that it credits the order of nature to the lowest possible level, the struggle among individuals (and even genes) for selective advantage. The framework makes some allowance for supplements to adaptation and graduated change, but restricts their scope. Gould's heresy is to enlarge that space, in the belief that evolution is richer and quirkier than current orthodoxy allows thanks to infusions of nonadaptive novelty, episodic jumps, and a heavy dose of sheer contingency. And Gould, Eldredge and their supporters are no longer lone voices in the wilderness. Their call for a more hierarchical view of nature, and for the restoration of the organism to its traditional place of honor, finds echoes in the writings of some developmental biologists and of students of complex systems.

* * *

Gilbert, Opitz and Raff, among others (10), make the argument for a broader and more robust framework from the embryological side. A renovated synthesis will be founded upon three basic concepts that were eclipsed by the redefinition of evolution from changes in organisms to changes in gene frequency, but have recently been rediscovered. The first is the return of macroevolution as a distinct field of study. Many biologists have come to agree that shifts in gene frequency (which nicely explain why the sickle-cell trait is so widespread in malarial regions of Africa) cannot account for the origin of hair or compound eyes, let alone the emergence of mammals from reptiles. No one argues that genes are irrelevant, but rather that such major transformations require novel genes and also drastic reorganization of the pathways of development.

The second rediscovery centers on the meaning of homology. Homology was crucial to the anatomists of the eighteenth and nineteenth centuries, who first realized that the vertebrae of mammals, birds and amphibians are in some sense "the same." With the rise of Darwinism, homology was recast as evidence for common descent (*analogous* structures, such as the wings of birds and insects, perform the same function but do not share a common ancestry). Common descent has, of course, not been abandoned. But we now appreciate just how deep those common roots go, and how severely they constrain the opportunities for change. Evolution, like politics, is an art of the possible. Change is permissible, but only so long as it does not disrupt the balanced workings of the whole organism. Structures and processes that carry a large burden of responsibility (e.g., ribosomes or vertebrae) can be modified but not replaced. We have long been aware that the core processes of biochemistry are universal, and must have been conserved from the very beginning of cellular life (Chapter 4). All the same, it came as a great surprise to learn that proteins that regulate cell division in humans and in yeast are so similar that the human version can complement a defect in the homologous yeast gene; and that homologous genes are required for the development of both the vertebrate eye and the insect eye (organs that used to be cited as the epitome of convergent evolution of analogous structures). No one is entirely certain what these recent discoveries are telling us, but one message is surely about a deep strain of conservatism beneath the exuberant diversity of the living world.

The third rediscovery restored to embryology what had long been its defining idea, the morphogenetic field. In Chapter 7, I deduced the need for this concept from reflections on growth and form in unicellular organisms; but its historical origin lies in animal embryology. Early in the twentieth century, embryologists became convinced that the keys to development lie hidden in those spatial territories within which all processes are coordinated by a common set of influences. No one has ever disproved the existence of such territories, but the idea was marginalized with the rise of developmental genetics, which substituted genes for fields as the instruments of spatial organization. The spectacular discovery of morphogen gradients and of homeotic genes has allowed embryologists to recover the concept. But the old bottle holds new wine: morphogenetic fields are now visualized as territories across which gene action is organized in space. The physical substratum of the various fields that operate in embryogenesis need not be everywhere the same. The universal principle is that developmental novelty results from changes at the field level, which thus underpin (animal) evolution. A high-level feature has been intercalated between the molecular gene and the whole organism: "Just as the cell is seen to be the unit of structure and function in the body—not the genes that act through it—so the morphogenetic field can be seen as a major unit of ontogenetic and phylogenetic change." (4)

*   *   *

To this point, the critique of evolutionary thought has remained under the umbrella of the modern synthesis. For Darwin and his followers, and even for most of their critics, natural selection remains the primary creative force, that generation upon generation favors the survival of the better adapted; adaptation, in turn, is the expression of a genetic program, honed and stabilized by selection among the phenotypes. For strict constructionists, organisms have no intrinsic form, organization or stability. They are mere figments of history, the ultimate in tinkerers' contraptions, and plastic as putty under the pounding of selection. One can imagine each organism perched upon a sharp peak in a fitness landscape, kept there by unremitting selection for better function, which weeds out the great majority of variants. Evolution becomes a mechanism for creating a high degree of improbability, but since these creatures are not governed by any laws of order, they remain at bottom unintelligible. This is not the universe inhabited by Brian Goodwin, Stuart Kauffman and others who approach evolution from the philosophical perspective called structuralism (11).

Goodwin draws historical inspiration, not from Darwin but from Immanuel Kant and the continental tradition of rational morphology, which dovetail with the new and still-evolving field of complexity studies. Certain complex systems, as we saw in Chapter 7, are naturally self-organizing; they generate spatial order from within, as a result of their own inherent dynamics. This holds for living organisms too: spatial order does not emerge from the genetic program, but from the dynamics that generate morphogenetic fields that, in turn, underlie morphology and physiology. It follows that biological forms are not fragile and contrived; quite the contrary, they are the "generic forms" most likely to be found by self-organizing dynamic systems, and therefore both probable and robust. Spatial patterns of this sort are famously fluid (think of flames or hurricanes), and are likely to vary in response to environmental conditions. We may imagine systems "exploring the space" available to the particular dynamics of each kind, and see evolution as the process by which their morphologies are transformed one into another. From this viewpoint, organisms are no longer artifices cobbled together by a capricious history, but the outward expressions of their own dynamics. Organisms recover their traditional status as fundamental entities, self-organizing and self-maintaining wholes. In principle, (and sometimes even in practice) they become intelligible because they are grounded in the laws of physics.

Genes and gene products do, of course, retain a role in the evolutionary drama. Catalysts and structural molecules determine the numerical parameters that enter into the physical specification of each system, and they stabilize its organization. Much of that exploration of the range of possible forms is, in fact, carried out by mutation and recombination of genes. But it is system dynamics, not the genetic program, that gives rise to biological forms and functions. By the same token, natural selection retains the role that has been described as "executioner of the unfit." But selection has been ejected from its throne as the dominant creator of biological form. Instead, the argument goes, spatial order is inherent in living systems, and that is all to the good. Stuart Kauffman, basing his ideas on extensive computer simulations of evolution, argues that selection by itself is too weak to maintain organisms atop steep and pointed fitness peaks. Evolution is possible only because selection collaborates with that spatial order intrinsic to living systems; and this order may persist ("shine through") even in the face of contrary pressure from natural selection.

In this brave new world, not only does evolution appear in a new light, so does phylogeny. Webster and Goodwin (11) do not deny the historical reality of shared ancestry and descent with modification. The green algae classified in the order *Dasycladales*, to which *Acetabularia* belongs (Chapter 7), are related as members of a family. But it is not their common history that accounts for the fact that the forms and functions of members of that order display variations on a common theme. Instead, Webster and Goodwin attribute it to the hierarchical nature of the processes that generate the organisms, the underlying fields in particular. "What emerges from this analysis is a theory of morphogenetic fields as complex dynamic systems that spontaneously undergo symmetry-breaking cascades: globally ordered initial fields pass through a series of bifurcations to detailed local structure that reflects initial order and results in an organism with coherently organized parts. The hierarchical nature of this generative process leads naturally to a hierarchical taxonomy of biological forms: ontogenesis provides the logical foundation for understanding phylogenesis" (11).

* * *

Where are we, then? Is the modern synthesis now an established doctrine immune to challenge, or is a radical reformation gathering speed?

Something in between, I think, and this continuing ferment is just what presently makes evolutionary biology such a lively province of science. The facts are not in dispute, but rather the view of life that flows from those facts; and there is latitude for personal perception. What seems to me to be happening is not unlike the trend emerging in molecular and cellular biology. Those disciplines have made enormous progress on the premise that dissecting cells into their constituent parts would enable us to understand the whole cell. That turned out to be true but less than all the truth, and we are now rediscovering the ways in which the higher levels of biological complexity control and constrain those below. By the same token, the modern synthesis is assuredly not incorrect. Common descent, adaptation by natural selection, the competition for survival, reproductive success, the production of variation by mutation and rearrangement of the genetic material explain many (even most) features of life. But they fail to explain why organisms exist in the first place, and they foster a splintered view of life that is at odds with the persistence and integrity of living forms.

Hence the appeal of that more generous conception, of which Stephen Jay Gould is presently the most doughty champion. A "pluralistic" Darwinism affirms the primacy of variation and selection, but allows greater play to additional forces. Among these are historical contingency, especially episodes of extinction caused by factors external to life, and hierarchical relationships among the nested layers of life, such that change and constraint can operate both upward from the level of genes and individuals, and downward from that of species and ecosystems. In the context of this book, the compelling merit of an enlarged synthesis will be the restoration of organisms to their traditional place as real entities in the hierarchy of nature. Living forms are not merely creatures of history, thrown together by accident and opportunity. Once again, organisms become natural kinds, the products of intrinsic generative procedures, malleable but not indefinitely plastic. Those who insist that the hierarchy of biological order must be compressed and pared down to its molecular foundations will, no doubt, read the foregoing as a surrender to mystification. I do not see it thus, but take it as a mark of progress that, a century and a half after Darwin, it is again becoming legitimate to see evolution as the historical adventure of *organisms* in geological time.

THE GENERATION OF NOVELTY

Cell evolution is a lot like human prehistory. It all happened long ago, the evidence is fragmentary and ambiguous, most of the links go missing, but opinions are strongly held and fiercely defended. The subject of this section is the emergence of cellular organization, the eukaryotic mode in particular, and our point of departure is once again the universal phylogenetic tree (Fig. 8.6). The objective now, however, is not to trace lines of descent but to highlight what the early stages of evolution suggest about the process itself. Cell evolution poses, in microcosm, the entire problem of how organisms came to be: self-maintaining and self-reproducing molecular systems, endlessly diverse yet united by common principles of operation.

To this end, it is instructive to view the advance of the microbes as a series of "inventions" which made possible the transition from one level of organization to the next (12). The first of these is that peculiar alliance of prokaryotes, which, if there is any merit to current opinion, initiated the branch that produced the eukaryotes (Chapter 8). The second transition, and by far the most spectacular, is that which generated the eukaryotic order itself, complete with something like the modern complement of features: genes arrayed on chromosomes and sprinkled with introns, a true nucleus bounded by a membrane inlaid with pores, an endomembrane system mediating vesicle traffic between the center and the periphery, and a cytoskeleton that provides mechanical support and integration. In the opinion of Thomas Cavalier–Smith, who has thought long and systematically, about this "abominable mystery," the key step in the transition was the invention of the cytoskeleton upon which all else hinges (13). The ancestral eukaryote, he believes, was a phagotrophic flagellated protozoan of the kind represented today by the Archezoa. It had no mitochondria, chloroplasts, peroxisomes or Golgi apparatus, but featured a single undulipodium with a sheaf of microtubules that sprang from its base and surrounded the nucleus. Mitosis took place without breakdown of the nuclear membrane. Energy metabolism was anaerobic, relying primarily on glycolysis. Secretory processes were primitive, but the organisms were motile and made a living by engulfing their bacterial prey. Protein synthesis relied on 70S ribosomes, as in bacteria, rather than the 80S ribosomes of the more advanced eukaryotic cells. (The metamonads of our own time fit this bill, but whether they can be taken to be "living fossils" surviving from extreme antiquity is in dispute; Chapter 8.) Cavalier-Smith envisages this huge amplification of cellular complexity as an autonomous process, building upon the elementary prokaryotic order, and has spelled out reasonable steps by which it may have come about. The extensive changes that this transition required may explain why it took some 2 billion years for eukaryotic cells to achieve their present prominence.

Other authors argue that even the first eukaryotic cell was the product of a partnership among prokaryotes, which supplied the precursors of the undulipodia and even of the cell nucleus. This view was originally held only by Lynn Margulis and her friends, but different versions of this idea are beginning to draw support from orthodox molecular biologists (Chapter 8). The two positions do not conflict as sharply as it seems, for both have much explaining to do as they strive to bridge the gulf between the eukaryotic mode of order and its prokaryotic precursors.

Many subsequent inventions are apparent in the proliferation of that ancestral flagellate's descendants, as they explored and adapted to new niches in an increasingly textured world. They all retained the eukaryotic mode of organization and its molecular building blocks, but they embellished the design and lengthened its reach. Symbiosis, and the gradual conversion of endosymbionts into organelles, clearly played a major role in the making of modern eukaryotes. Other innovations are less conspicuous. Amoeboid locomotion is one, based on the cyclic assembly and disassembly of the cortical cytoskeleton; it is the foundation of all crawling movements, from amoebas to human embryonic cells and apical growth in fungal hyphae. The secretory machinery, rudimentary in the ancestral enkaryotic flagellate, burgeoned with the invention of the Golgi apparatus. This elaborate switching station is clearly essential to the life of green algae and higher plants with their complex polysaccharide walls. Mitosis was transformed: "closed" mitosis persists in some protists, but in the majority of modern eukaryotes the nuclear membrane breaks down and is reconstituted after chromosome segregation. In some protists, structures made up of microtubule bundles came to serve as a rigid cytoskeleton, framing the gullet or the entire cell. Undulipodia, too, blossomed into strange forms. They are still the common mode of cell propulsion; but many protists now sport four undulipodia, one of which has been modified so as to drive food particles into the gullet. Gullets themselves come in diverse shapes, sizes and modes of operation, and so do all other features and facets.

Whence comes this blizzard of novelty, and how is the new integrated with what came before? Molecular phylogeny allows us to map the descent and diversification of genes and proteins, and casts light upon the genealogy of the organisms that bear those macromolecules. It is much more difficult to work out the genesis of complicated and integrated systems, such as undulipodia made up of over 200 proteins working in unison. As matters presently stand, novelty arises from three sources: gene duplication and divergence, symbiosis and epigenesis (12). Let us examine these in turn.

*   *   *

The generation of novelty at the level of genes and gene products is the result of unavoidable errors in DNA replication, and its molecular basis is generally well understood. Point mutations, recombination and other kinds of genetic rearrangement contribute to the march of mishaps, but the chief source is likely to be transposons, virus-like genetic elements that can jump from one place in the genome to another. For any kind of mutation to provide the raw material of evolution, the gene in question must exist in at least two copies so that one copy can continue to support the gene's original function, while the other is free to vary. Here again, the errors of replication that give rise to duplication are known, at least in principle. We have recently come to recognize that bacterial genomes, and perhaps others, encode the capacity to generate particular kinds of mutants on a massive scale, in response to dire emergencies such as starvation. In general, genomes turn out to be much more plastic than was believed until recently, thanks to enzymes that cut, splice and rearrange coding sequences in response to environmental signals (14). Sources of variation abound; by way of compensation, all organisms also possess an elaborate enzymic machinery to ensure the accuracy of gene replication and to keep the basal incidence of mutations in any one gene around one in a million for each cycle of replication.

Despite such careful "proofreading," when one examines populations of organisms, one finds a remarkable amount of variation at the level of genes and proteins. Most macromolecules come as assemblies of closely related sequences that differ at one or two positions but are functionally indistinguishable. Even individuals may carry several such isoforms. Neutral mutations of one kind or another make up the lion's share of molecular diversity, and they raise the question whether variation is necessarily adaptive. Mutations that have no functional consequences are invisible to natural selection; neutral mutations probably accumulate in populations, not because selection favors them, but because it cannot keep them down. Molecular isoforms may, however, become the starting point for the evolution of novel functions. Genes often come in families (genes that code for transport carriers, for instance) that arose by the duplication of an ancestral gene followed by the gradual divergence of the copies. One cannot help wondering whether rearrangements on a grander scale, either of genomes or of cell structure, can provide points of departure for the origin of species or even or higher taxa.

The other face of the Darwinian coin displays natural selection. New or altered genes can become fixed in populations by random events ("genetic drift"), particularly when the gene product does not carry a major functional burden and the population in which the mutation occurred is small. But the consensus among evolutionary biologists holds that systematic changes, of the sort that alter form or function and promote adaptation to changes in the environment, are invariably the result of natural selection. Selection serves, first and foremost, as a stabilizing force by weeding out mutations that diminish fitness. But it also favors the propagation and spread of the rare variants that enhance adaptation to the circumstances that prevail at the moment. Selection is for the here and now; it has no foresight, and cannot anticipate what functions may be useful in the future.

With the focus of contemporary research fixed on the molecular level, it is tempting to identify the "great moments in evolution" with the appearance of key molecules (15). Actin, tubulin, calmodulin and the lamins of the nuclear matrix certainly qualify, and so do many others. Such proteins are ubiquitous among eukaryotes, and they have been strikingly conserved over a billion years of chance and change. Significantly, each of these proteins serves as a building block utilized in a range of cellular functions, as bricks do in construction. But tubulin alone does not an undulipodium make, just as bricks do not suffice to make a house. To achieve a working machine, or a pathway that can carry signals from the cell periphery to the nucleus, the key proteins must be linked to others in a purposeful manner. Much of cell evolution turns on these links: G-proteins, receptors, actin-binding proteins and motor proteins. These are typically more variable than the key proteins; ' while the core elements are conserved, new functions arise, as it were, by incremental tinkering with the molecular periphery.

The fundamental postulate is that unduplipodia and other multi-molecular mechanisms arose, like the human eye, by the progressive accretion of ancillary proteins onto some rudiment or foundation that was functionally useful but need not have been an organ of motility. This amplification took place, one gene at a time, under the guidance of natural selection: each modification conferred at least a small selective benefit. On this premise, one can construct schemes that sound plausible and account, in principle, for the origins of crawling motility, mitosis or the secretory pathway (15). We have no better alternative to offer the inquirer, and in the absence of time travel we may never discover what actually happened; and so a modicum of doubt necessarily persists. We should reject, as a matter of principle, the substitution of intelligent design for the dialogue of chance and necessity (16); but we must concede that there are presently no detailed Darwinian accounts of the evolution of any biochemical or cellular system, only a variety of wishful speculations. There is room for discovery here, and for reflection too; nowhere is the appeal of Gould's "pluralistic Darwinism" more keenly felt than in the study of cell evolution.

*   *   *

Genes did not necessarily arise where they are found, they may have been imported. There is something disturbing about the notion that genes can be transferred from one species to another, let alone between phyla and kingdoms. It violates one's sense of organismal integrity, and calls into question the principle of a lineage defined by the vertical transmission of genes from parent to offspring. But horizontal transfer happens, on a scale that ranges from single genes to entire genomes; it represents a major source of evolutionary novelty, and a significant enlargement of the modern synthesis.

One such kind of lateral gene transfer has become painfully familiar: the spread of antibiotic resistance among pathogenic bacteria. Resistant mutants have been known since the advent of the antibiotic era half a century ago; what makes them a menace is the propensity of genes that confer resistance to cluster on plasmids, virus-like strings of DNA independent of cellular chromosomes, that pass easily from one bacterial tribe to another. The widespread use of antibiotics in both hospitals and animal feed has created an environment that promotes the spread of resistance, and the use of antibiotic resistance as a marker in the genetic modification of crop plants seems calculated to make matters worse.

How widespread in nature is horizontal gene transfer? Among bacteria it may be very common indeed, thanks to viruses, plasmids and other mobile gene carriers (17). A recent examination of the *E. coli* genome suggests that out of 4,288 protein-coding genes, 755 or 18 percent have come from other organisms since *E. coli* diverged from *Salmonella* 100 million years ago. Among that number are the genes that make some *E. coli* strains pathogenic. Genes are also known to pass between prokaryotes and eukaryotes. *Agrobacteria*, plant pathogens that elicit tumor formation, are thought to have transferred some of the pertinent genes to their plant hosts. Conversely, at least one of the key enzymes of glycolysis in *E. coli* now appears to be of very ancient animal origins. Certain long-standing controversies over the taxonomy of flowering plants can be resolved by invoking extensive exchange of genes between lineages. And massive gene flow at the very dawn of bacterial evolution (dubbed "genetic annealing" by C. R. Woese) has been invoked in order to explain the unexpected finding that Archaea share some genes with the Eukarya but many more with the Eubacteria (Chapter 8). All the same, one comes away with the impression that horizontal gene transfer is presently an infrequent, even marginal phenomenon that does not seriously blur the lines of phylogeny.

The view alters markedly when we turn from the lateral transfer of single genes to the wholesale melding of lineages. The term "symbiosis" covers a spectrum of associations whose common feature is that they bring together in varying degrees of intimacy genomes from organisms that are effectively unrelated. In the absence of symbiosis, the world would be a very different place (18). There would be no lichens to clothe bare rocks, no mycorrhizal fungi to nourish the roots of forest trees, no termites to feast on wood with the help of cellulase secreted by the protists that inhabit their hindguts. There would be no oases of life around deep-sea vents, for the clams and tubeworms that flourish there have neither mouth nor anus and rely for their nutritional needs upon symbiotic bacteria that oxidize the abundant hydrogen sulfide. No stable symbiosis has ever been observed among prokaryotes, but they are common among eukaryotes of all kinds; this suggests that symbiosis is critically dependent upon some feature unique to eukaryotes, and that feature is likely to be the capacity to engulf prey and other particles.

Symbioses in which one partner takes up residence in the cytoplasm of the other (called endosymbiosis in the literature) are common among protists, and while some look distinctly bizarre to our eyes, they ring with echoes of biological antiquity (19). Certain unicellular green algae lack chloroplasts, instead, they harbor organelles called cyanelles, now known to descend from endosymbiotic cyanobacteria. Cyanelles retain the peptidoglycan cell wall of their bacterial ancestors, but can no longer live independently because most of their genetic complement has been transferred to the algal host. It is irresistible to see cyanelles as chloroplasts in the making, caught in transition. A report from Copenhagen describes a "purple protist," an anaerobic ciliate that harbors photosynthetic bacteria as endosymbionts and can grow under strictly anaerobic conditions in the light; could this be a model for the origin of mitochondria? From Groningen comes news of another anaerobic ciliate, a denizen of the cockroach hindgut, which hosts methanogenic archaebacteria; it also has a hydrogenosome that retains a genome with mitochondrial affinities, clear support for the hypothesis that hydrogenosomes derive (directly or indirectly) from bacterial symbionts. And then we have that champion of multiculturalism, *Mixotricha paradoxa*, which lives in the hindgut of termites. It hosts at least four kinds of bacterial endosymbionts, including spirochaetes attached to the surface which serve the protistan cell as organs of motility. Does this creature hold a clue to the origin of undulipodia, or does it merely sow confusion?

Such odd partnerships do not necessarily recapitulate the phylogeny of organelles, but they do show that the era of evolution by association is not over. One such episode has even been captured in the laboratory. In the course of his researches on amoebas, K. W. Jeon (19) noted that his organisms were sickly due to a bacterial infection. The strain survived, not by expelling the bacteria but by accommodating to them. Over a span of two hundred generations, host and parasite became dependent on each other so that neither can now grow in isolation (the reason for their interdependence remains unknown). It looks like another early step along the road whose end-stage is a full-fledged organelle.

Does evolution by symbiosis contradict the dogma of gradual change by small steps? Well, yes and no. Those purple protists acquired their photosynthetic powers by a sort of saltation, a "punctuational" event. But the photosynthetic apparatus of the bacterial symbionts was surely honed by the orthodox interplay of variation and selection. We should also consider once more whether every step in evolution is guided by natural selection; this is clearly not the case for the early stages in the evolution of Jeon's amoebas, but true for the later ones. In dense microbial communities predation and parasitism are both prevalent, with the occasional emergence of a novel combination that may be available for selection as a new phenotype. Perhaps the most disconcerting aspect of evolution by symbiosis is that mainstream evolutionists remain so reluctant to incorporate it into their historical vision.

* * *

"Epigenesis," like many other terms of scholarship, means precisely what the speaker wants it to mean. Historically, epigenesis supplied the counterpoint to preformationism; it stood for the commonsense view that embryonic structures arise in the course of development, rather than being present beforehand in egg or sperm. In Chapter 7, I used the term broadly to designate all the processes by which genetic information (modified by influences from the internal and external environments) is translated into the substance and behavior of an organism. Geneticists prefer to restrict "epigenetic" to those heritable changes in gene expression that do not involve alterations in the sequence of nucleotides. Whatever definition best serves the purpose at hand, one can scarcely doubt the importance of epigenetic processes in physiology, morphogenesis and development. The question is whether epigenetics also serves as a source of evolutionary novelty. For this to be feasible, some of the structures and functions that arise during development must be both variable and heritable, over and above the information encoded in the genes, and that proposition continues to be regarded with deep suspicion. It calls up Lamarck's baleful ghost, implies that acquired characteristics may be inherited after all, and seems to threaten our hard-won understanding of both heredity and evolution.

In fact, as Eva Jablonka and Marion Lamb explain in their lucid and thoughtful book (20), the barriers of principle were breached long ago. In animals, it is true, changes that arise during development or in adults are seldom inherited. The reason is that animals possess a discrete germ line that arises early in development; only alterations that survive meiosis can make it into sperm or egg and be passed on to the next generation. But this restriction does not apply to organisms whose somatic cells can reproduce, and epigenetic inheritance is well known from the world of plants and particularly that of microorganisms. Epigenetic inheritance comes in a variety of guises (20). Some fall within the purview of the molecular biologist, particularly covalent modifications of certain DNA bases that alter the rate of transcription. The methylation of promoter sites is a case in point: as a rule, promoters of genes that are available for transcription are unmethylated, while methylated ones are silent. The pattern of methylation persists through cell reproduction, thanks to an enzyme that recognizes DNA molecules in which only one strand bears methyl groups and makes up the deficiency. This is one of the mechanisms that ensures that liver cells express genes for liver functions but not those for kidney function (and *vice versa*), even though both cell types possess the entire gene complement.

Other examples of epigenetic inheritance come from cell physiology, such as the persistence of certain metabolic states from one generation to the next. In *E. coli*, the enzymes for lactose metabolism are only expressed when lactose is present in the medium; but once the active state has been established it can be propagated for many generations even in the absence of lactose. Most dramatically, we saw in Chapters 6 and 7 that the transmission of cellular structure between generations is partially independent of the genes. Jablonka and Lamb cite Boris

Ephrussi, who saw long ago that "not everything that is inherited is genetic." Cell membranes, it seems, are inherited from one generation to the next, but the implications of this continuity have never been thoroughly explored. It seems beyond dispute that cells possess two systems of inheritance, one genetic and the other epigenetic. The question remains whether epigenetic inheritance is stable enough and common enough to make a significant contribution to cell evolution.

The proposition that epigenetic inheritance can be a source of evolutionary novelty is especially inviting (albeit altogether speculative) as a basis for reflection on the manner in which the multitude of unicellular forms may have come about. We saw in Chapter 7 that form is not directly or rigidly determined by the genotype: the genes define a range within which the phenotype falls, but forms arise epigenetically as the result of developmental processes. In the genesis of this superior level of order, morphogenetic fields appear to play a crucial role, and they are commonly propagated through structural inheritance. Variations are likely to be frequent. Most of them, like most genetic variations, will be detrimental, but on rare occasions one may give rise to a "hopeful monster" that can gain access to a novel way of life. The new order, if at all functional, would become subject to natural selection as a new unit: mutation and selection hone the genes that stabilize the new organization and improve its operation. Adaptation would occur by orthodox Darwinism processes, but the novelty would have arisen by a nonadaptive leap.

This, I suppose, is what Cavalier-Smith meant when he suggested that the seminal events in protistan evolution, those which resulted in a new kind of organism, should be seen as "internally generated accidents rather than adaptive responses to external conditions" (13, 1991). For example, he believes that the divergence of the Eukarya began when their bacterial ancestor lost the capacity for cell wall synthesis, and the crippled offspring evolved alternative means to segregate the genome at division. (L-forms, cell lines that have lost the capacity for cell wall synthesis but continue to propagate all the same, are well known to bacteriologists).

Do hopeful monsters and evolution by saltation ever happen in the real world, or are they but figments of an overheated imagination? We don't know, and it is not obvious what kind of evidence would either verify or refute the proposition. There are plausible candidates among the protists. Joseph Frankel considered one such, a species of protozoa that may have arisen by reorganization of the cell cortex of another species. Among the green algae one finds morphological diversity that could represent structural variants. Few scientists now study these organisms, and in the absence of evidence one can easily deride structural saltation as so many "just so" stories. But Gilbert et al. (4) put their fingers on the key point: "Evolution depends on the replication and modification of morphogenetic fields. . . . The mechanism by which fields can be replicated and then altered is a new area of research which should produce new insights into the mechanisms of evolution." Amen.

A SINGULARLY SUCCESSFUL DESIGN

This much is certain: the eukaryotic mode of cellular organization has been singularly successful. It has ramified into some 200,000 unicellular species, and laid a firm foundation for the evolution of multicellular forms—including one with the capacity to ponder evolution. Nothing seems to forbid alternative patterns that exceed the prokaryotic grade in complexity and versatility, but if such ever existed they failed to thrive, nor have intermediates between prokaryotes and eukaryotes survived. One has to wonder why this is the case: what is so special about the eukaryotic design that allowed it to flourish and multiply exceedingly well? And is such excellence itself a product of natural selection?

For a start, we can ask about the place of eukaryotes in the economy of nature, and how it differs from that of prokaryotes. Protists and bacteria share waters, soil and odd corners inside higher organisms, but they pursue different strategies. These, as Michael Carlile pointed out (21), relate to the disparity in size and organization. Bacteria are minute, ubiquitous, omnivorous, and quick to multiply. This favors bacteria where nutrients are abundant but transient: the cowpat dropped into a puddle soon hosts a goodly crop of bacteria. Protistan cells are larger, have lower metabolic rates and multiply more slowly. They lack the biochemical versatility of bacteria, but can exploit their small competitors by preying on them or by domesticating them as endosymbionts and organelles. Thanks to their capacity for photosynthesis eukaryotes excel as the primary producers of organic matter, and they generally predominate in aerobic environments where resources are scarce but relatively stable. Here the race to leave the most progeny will be won, not by those quickest to reproduce but by those who are best adapted to a demanding home range. It is all too easy to point out exceptions to this thesis, but there is a ring of truth to the notion that eukaryotes have adopted novel strategies—not economic, perhaps, but evolutionary.

The Eukarya have a history of repeated radiation on multiple levels—diverse kinds of protists, fungi, plants and animals—and each radiation created additional niches for eukaryotes (inside insects, for instance, whose hindgut shelters symbiotic protists). Obviously, the eukaryotic pattern has whatever it takes to produce variations that are themselves successful. This quality is referred to as "evolvability," recently defined by Marc Kirschner and John Gerhart (22) as "the capacity to generate heritable, selectable phenotypic variation." Note that there is more to variation than making mutants. Natural selection sees phenotypes, not genotypes, and in consequence the evolutionary prospects of any mutation hinge on the cellular context in which the genetic information is expressed. What is it, then, that makes eukaryotes so evolvable? Kirschner and Gerhart look to the physiological and developmental processes that mediate between genotype and phenotype. From the standpoint of evolution, the flexibility and tolerance of these epigenetic activities is not mere noise. On the contrary, developmental plasticity widens the range of phenotypes, providing a richer assortment of variants on which selection can act.

Kirschner and Gerhart draw their examples from the higher animals, but the proposition that evolution is facilitated by a cellular design that loosens the linkage between genotype and phenotype applies to eukaryotes generally. They point out that the key building blocks have been strongly conserved (actin, tubulin, calmodulin among others); organisms diverge by rearrangement of the regulatory circuitry. Regulation is typically mediated by proteins that are relatively unspecialized and can be fitted for new functions with a small number of mutational changes. New functions can also arise by accretion, as gene transcription illustrates. In prokaryotes, the proteins that control a given gene are few in number and highly specific; regulation is tight. By contrast, in eukaryotes gene transcription is controlled by multiple proteins that interact weakly with DNA and with each other; new proteins bearing additional messages can be added on and summed up with the prior ones.

This sort of fluidity also shows up in cellular construction. For example, take the mitotic spindle. A chromosome can move only when attached to a spindle fiber, but there is no map or blueprint that guides the fibers to their target. On the contrary, spindle fibers are nucleated by the thousands, sprouting and shrinking continually. But when, by chance, a fiber contacts a chromosome at the proper locus the fiber is stabilized and can now exert traction. The process is wasteful, expensive and redundant, but it is also robust: a working spindle can organize itself with a minimum of specification and from many initial configurations. Finally, cellular organization (both functional and structural) is characteristically modular, made up of standard parts. The virtues of a hierarchical organization consisting of interconnected but discrete parts were spelled out by the late Herbert Simon four decades ago (23). Modules need not be precisely tailored to one another, but can usually function in diverse contexts (wheels, for instance); and the organization as a whole is less likely to be disrupted by a localized failure. In the evolutionary context, the point is that modular organization increases flexibility and minimizes cascading malfunction. These and other features of eukaryotic cell biology act cooperatively to make their lineage "evolvable" by diminishing the damage associated with mutation, and by making available more favorable and nonlethal kinds of variation for selection to cull.

Is evolvable design itself a product of natural selection? That is a fraught question, for in the neo-Darwinian schema evolution results chiefly (if not solely) from competition among individuals for immediate reproductive advantage. Selection for evolvability, by contrast, would be exerted upon the whole lineage ("clade selection") and confers future rather than present benefits. One way to accommodate evolvability in the orthodox framework is to argue that flexible coupling between phenotype and genotype makes for more robust physiology, and thus enhances the fitness of individuals; evolvability of the lineage is not the target of selection but a spectacular byproduct (a spandrel, if you will). But direct clade selection remains in the running, because flexible coupling allowed eukaryotes to radiate quickly into new or vacant niches to the exclusion of competitors. In fact, the history of early life is liberally sprinkled with major episodes of extinction and radiation (24). I find it stimulating to imagine that the capacity for evolutionary change itself evolved over time, gathering speed as one layer of flexibility was superposed upon another. Eventually there emerged, in a single lineage, the most potent and adaptable innovation of all, language; and the character of the entire biosphere was transformed for evermore.

A DIRECTION TO EVOLUTION?

There is something rather comforting about the great tree of life, an intimation of continuous advance and even a whisper of ultimate meaning. Beginning with shadowy primordial forms, evolution brought forth first the small and relatively simple prokaryotes and later the larger and more sophisticated eukaryotic protists. In due course, the latter gave rise to a prodigious flowering of organisms, both unicellular and multicellular, aptly designated the crown of the eukaryotic stem; it culminates in the emergence of the higher plants and animals. Another and yet more advanced stage was reached when a single and unique species of primates, endowed with a greatly enlarged brain, achieved speech, consciousness and the capacity for reason and reflection. The tree encourages one to hope that, before too late, our minds can transcend our animal instincts, making us stewards of our heritage rather than its consumers.

It seems perverse to deny the obvious signposts of progress (defined in my *Collins English Dictionary* as "advance towards completion, maturity or perfection"); but it is surely the task of scientists and philosophers to look behind the appearances, and so the direction of evolution and the parade of progress have become matters for lively debate. Darwin himself seems to have been of two minds. He could hardly ignore the fossil record, whose mounting complexity supplied the evidence for descent with modification, but he insisted repeatedly that natural selection, which allows organisms to adapt to changes in the local environment, provides no reason to expect a global advance in complexity or perfection. Contemporary biologists are sharply divided. Stephen Jay Gould, reviewing the issue in his recent book *Full House* (25) quotes his equally eminent colleague E. O. Wilson: "Progress, then, is a property of the evolution of life as a whole by almost any conceivable intuitive standard, including the acquisition of goals and intentions in the behavior of animals. It makes little sense to judge it irrelevant." Gould himself comes down squarely on the opposite side: he "fervently" denies that the evolutionary record supports "an argument for general progress as a defining thrust in life's history." To be sure, over time the record dramatically displays the successive emergence of increasingly complex organisms. But the appearance of progress results from a history of expanding variation away from a "fixed wall" of minimal sophistication. The direction of mounting complexity is the only one open to invasion, and there is no evidence whatsoever that complex forms are inherently superior to simpler ones, or that they are consistently favored by natural selection. In fact, Gould argues, it is we humans who read progress into nature for our own reassurance: "We crave progress as our best hope of retaining human arrogance in an evolutionary world."

Well now. Let us, for starters, eschew that loaded word "progress," and speak of complexity instead. That term is not easily pinned down either, but most folks share an intuitive sense that complexity turns on the number and variety of interacting parts. The complexity of one cell species relative to another increases in proportion to the number of constituent molecules and the number of different kinds; it increases with the density of molecular interactions, and with the number of genes that code for some function. Structural intricacy enters as well, but is harder to assess. In higher organisms the number of cell types makes a useful measure: 3 in yeast, about 250 in humans. In ecological communities, the roster of species is an index of complexity. By all these criteria, the history of life displays a remarkable progression of complexity over time—not uniformly or universally, but so generally as to demand explanation. Is there some global force that drives organisms (or the biota as a whole) in the direction of greater complexity? Does natural selection tend to favor more complex organisms over simpler ones? Or is the trend, as Gould believes, merely an effect of evolution's random walk that generates an increasingly skewed distribution of living forms?

The proposition that evolution proceeds in an oriented fashion toward ever greater complexity is very widely held and seems virtually self evident, but Gould finds no evidence to support it. Today, as throughout evolutionary history, the most abundant organisms by far are also the simplest—bacteria in particular. Advanced organisms make up but a minority of the total, the tail of a highly skewed distribution; there is no sign of any overall ("modal") increase in size or complexity of the biota as a whole. Any directional trend should show up in the fossil record of particular lineages, and there are classic instances that have been so interpreted: the steady progression of horses toward a single toe, the growing complexity of the sutures of ammonites or of the vertebral column from fish to humans, the tendency of foraminifers to become larger over time. But not one of these stands up to critical scrutiny. Even when the trend is real, it is but a statistical consequence of increasing variation, and there is no indication that greater complexity contributes to the abundance or longevity of species. The rise of the higher organisms is nothing more than an incidental consequence of the steady proliferation of living forms, each of which must necessarily start from that fixed wall of minimal complexity. Repeatedly Gould likens the spectacle to a drunk staggering blindly from the pub's wall to the gutter; an unfortunate choice of imagery, for it quite fails to illuminate that unsteady but undeniable ascent of "Mount Improbable" that so impresses everyone else.

From the perspective of genetics and physiology, the sense of mounting complexity, versatility and autonomy is so pervasive that something more than pure drift really must be invoked. The advances are neither universal nor inevitable (at every level, some lineages buck the trend), and they supply no evidence for an external driving force or for some inherent tendency to elaboration. The ultimate cause of the phenomenon must, therefore, be sought in natural selection, as John T. Bonner,

John Maynard Smith and Eörs Szathmàry have all done (12). It is helpful here to keep in mind the deep cleavage that separates the two modes of biological organization, prokaryotic and eukaryotic. Bacteria, though biochemically both diverse and adaptable, have remained structurally simple and developmentally limited. The constraints may fundamentally be a matter of small genome size, itself a product of selection for rapid growth and reproduction. When is comes to generating genetic and structural complexity, eukaryotes draw the eye. Some, it is true, have become simplified over time, in the way that parasites shed unneeded capacities and appendages. But what stands out is that sequence of transitions (molecular, anatomical, genetic) that opened the door to new opportunities: actin and tubulin, chromosomes, complex cellular architecture, sex, multi-cellularity, language. . . .

As far as one can tell from the record, natural selection does not favor increased complexity *per se*. But many of the inventions and transitions that selection has favored did, in fact, make organisms more complex. This general bias should not be surprising, for innovation commonly comes about by accretion. New genetic information originates in gene duplication, followed by divergence; genes may be acquired by lateral transfer, or *en bloc* by symbiosis; and new layers of command and control impose spatial and temporal order on the expression of that information. We should expect systems that evolve in this manner to throw up forms of increasing complexity, some of which are bound to succeed. Any planet that hosts life will therefore feature a full house, life in abundance and diversity woven into a tapestry that mingles the simple and the complex.

Thanks to Darwin and his successors, the plenitude of life is no longer a mystery. But it remains a marvel beyond compare, a spectacle of surpassing grandeur and never-ending fascination. Aldo Leopold in his *Sand County Almanac* (26) drew from it a basic principle to guide public policy: "A thing is right when it tends to preserve the integrity, stability and beauty of the biotic community. It is wrong when it tends otherwise." Whenever we allow the clamor of commerce to drown out that quiet voice of reason and restraint, the world is the less; and every species lost by our own actions diminishes us.

10

So What Is Life?

"Now one could say, at the risk of some superficiality, that there exist principally two types of scientists. The ones, and they are rare, wish to *understand* the world, to know nature; the others, far more frequent, wish to *explain* it. The first are searching for truth, often with the knowledge that they will not attain it; the second strive for plausibility, for the achievement of an intellectually consistent, and hence successful, view of the world."

Erwin Chargaff (1)

An Ingenious Machine?
Complex Systems
Living by the Second Law
A Unicorn on the Wall

Fifty years after Schrödinger wrote his little book, his challenge still hangs in the air. What *is* life? Having learned so much about molecules and mechanisms, structures and functions, physiology and ecology and ontogeny and phylogeny, why are we still at a loss for a satisfying answer? Schrödinger himself posed the riddle with a flourish, but wisely refrained from offering a solution; today we are quick to deflect the mystery with a wry smile, a parable, or a joke (2). The reason has much to do with the difference between explanation and understanding. We are quickly learning to explain the workings of the biological machinery and even how organisms came to be as we find them, but we have no persuasive answer to the question why life exists in the first place. Loren Eiseley, thirty years ago, was baffled by "the hunger of the elements to become life," and we are not much wiser today. There is nothing in the textbooks of physics and chemistry to forbid a world that teems with bacteria and butterflies, but there is also nothing that would lead one to expect the world to be of this nature. The crux of the matter is that living organisms cannot be rationally and systematically deduced from the principles that generally do account for the properties of inanimate matter.

We biologists claim for our science a high degree of autonomy from chemistry and physics, and rightly so. Organisms are historical creatures, the products of evolution; we should not expect to deduce all their properties from universal laws. The antics of a troop of monkeys in the forest canopy are doubtless consistent with all of physics and chemistry, but this knowledge supplies no insights that will be useful to a student of animal behavior. All the same, the autonomy of biology must ultimately trouble those who, with the late Jacob Bronowski, "seek to find nature one, a coherent unity" (3). The reason that many thoughtful persons continue to find life perplexing, even mysterious, is that sharp division between the organic and inorganic spheres (Chapter 2). The distinction turns on those characteristics that are universally associated with entities we designate as living, but essentially absent from nonliving ones: intricate organization and purposeful behavior that unfold over time, both on the individual level and that of the total assemblage. Here yawns a great chasm that all biological scientists recognize, but many are deeply reluctant to acknowledge.

There is clearly something special about living things that has not declared itself from beneath our vast heap of knowledge, and that seems to stand outside the circle of light that contemporary research strives to enlarge. What we lack is an understanding of the principles that ultimately make living organisms living, and in their absence we cannot hope to integrate the phenomenon of life into the familiar framework of physical law. I am not here to advocate a veiled vitalism, nor to sneak in a creator by the back door. But I do insist that until we have forged rational links between the several domains of science, our understanding of life will remain incomplete and even superficial. Until that impasse is overcome, we cannot refute philosophers, skeptics, religious believers and mystics who suspect that science is sweeping out of sight the very mystery that it purports to elucidate.

I do not have the answer to Schrödinger's riddle; no one does. It is even conceivable that we stand here at one of the limits of science, but it would be quite premature to concede defeat. We are gravely hampered by having but a single kind of life to ponder, and it may turn out that we cannot fully grasp the general phenomenon until we have either found additional versions of life or produced one in the laboratory.

Neither prospect seems bright at present. In the meantime, however, we can wander a little down earthly paths that may, perhaps, offer glimpses of that elusive consilience between the physical world and the biological one.

AN INGENIOUS MACHINE?

We shape our buildings, Winston Churchill once said, and then they shape us. In just this fashion, the questions we ask about life and what we take for answers have been molded by two figures of speech: first the metaphor of the machine, and latterly that of the computer.

Wander the corridors of a biology building to eavesdrop on the residents, and you'll find them preoccupied with nuts and bolts. We argue about the mechanism of photosynthesis, of embryonic development or even of evolution. We speak earnestly of the "machinery of life," of biological "building blocks," of traits "hard-wired" in the "blueprint"; and we prepare students for careers in the new manipulative science, particularly in "genetic engineering." Even medicine, that most humane of arts, seems to be turning into a science of spare parts, plastic or molecular. Isak Dinesen caught the mood early (4): "What is man, when you come to think upon him, but a minutely set, ingenious machine for turning with infinite artfulness the red wine of Shiraz into urine."

A machine, Webster informs us, is "an assembly of parts that transmit force, motion and energy to one another in some predetermined manner and to some desired end." It is a stretch, but seemingly not an excessive one, to apply this definition to living things: chemical machines, whose object is to make two where there was one before. The metaphor carries immense heuristic power. It conjures up a system of interacting parts with well-defined functions, joined together in service to a greater whole; of energy supporting useful work, and of powers emerging in the machine as a whole that were not present in the isolated parts. Every time we announce the unraveling of another mechanistic puzzle, we pay tribute to the metaphor of the machine: mechanisms are a real feature of living beings. Besides, the metaphor resonates with our deepest conceptions about the way things are. For three hundred years now, scientists have perceived the world through the eyes of Descartes and especially of Newton: a universe of particles moving in fields of force, whose behavior is fully determined by the overarching laws of physics. For those who subscribe to this viewpoint, biology is little more than a collection of special cases that must be accommodated within the general framework; no deep mysteries there.

Well then, what is wrong with the assertion that a cell—*E. coli*, say, or a ciliate—is just a particularly intricate and ingenious machine? The fault is that the claim begs the central issue. If a cell is just another machine, what is the basis for the distinction that has been drawn from ancient times between objects that are alive and those that are not? After all, what we seek to understand is not what these two categories have in common, but what sets them apart! The answer came in the eighteenth century from the German philosopher Immanuel Kant (5), and turns on the existence of a special category of objects called organisms. In a machine, Kant said, the parts exist for each other but not by each other; they work together to accomplish the machine's purpose, but their operation has nothing to do with building the machine. It is quite otherwise with organisms, whose parts not only work together but also produce the organism and all its parts. Each part is at once cause and effect, a means and an end. In consequence, while a machine implies a machine maker, an organism is a self-organizing entity. Unlike machines, which reflect their maker's intentions, organisms are "natural purposes." Kant's vision was eminently sensible and remains true, but even he was stymied by the next stage: How can we ever discover the cause of that purposeful organization that is the hallmark of organisms?

Time and the advance of science have supplied a partial solution to that enigma, and produced a more nuanced view of the relationship between organisms and machines. To Kant's distinction between machines that are made and organisms that make themselves, we would add that machines can be built singly as the need arises; some are unique. Organisms, by contrast, invariably occur as members of an ecosystem, a historical community whose transformation over time molds the forms and functions of its members. Most contemporary scientists, if they give any thought at all to such abstruse matters, will subscribe to a conception that incorporates elements of both organism and machine, and which is grounded in our new understanding of heredity, biochemistry and evolution. The metaphor of the day, not surprisingly, describes organisms as information-processing devices, computers of sorts. The image leaps to the eye: behold the DNA tape whose nucleotide sequence stores information that can be replicated, read out and expressed in the language of proteins, those minuscule devices that do all the work. Thanks to rare but unavoidable errors the tape is altered over time, generating new versions for natural selection to sift. If organisms are still machines they are exceedingly artful ones, that do not fit easily into the traditional understanding of mechanical appliances.

For all its contemporary tone, the metaphor of the computer has substantial historical depth. Webster and Goodwin (5) trace it back to

August Weismann who, at the end of the nineteenth century, drew a sharp line between the germ line and the soma. In place of Kant's self-organizing whole, Weismann preached a fundamental duality, which pervades our thinking to this day. The genome, continuous from the very dawn of time and in principle immortal, serves as the central directing agency for the production of the visible animal or plant. These bodies, made by executing the instructions contained in the genome, are effectively artifacts; they mediate between the genome and the environment (to borrow a phrase from Jeffrey Wicken), but have no intrinsic significance. Details aside, there is no denying that this way of looking at life has a tremendous hold on all our imaginations. One cannot help wondering, as Webster and Goodwin do, how much that grip owes to the far more ancient dualities of body and soul, matter and spirit.

The philosopher Alfred North Whitehead put this in perspective: There are no whole truths, he said, all truths are half-truths; the trouble comes from treating them as whole truths. The informational metaphor is assuredly a substantial half-truth but still only that, and readers who have trudged through the preceding six chapters will see why. Organisms process matter and energy as well as information; each represents a dynamic node in a whirlpool of several currents, and self-reproduction is a property of the collective, not of genes. Form, structure and function are not straightforward expression of the gene's dictates; there is more to heredity than what is encoded, and you can only go from genotype to phenotype by way of epigenetics. DNA is a peculiar sort of software, that can only be correctly interpreted by its own unique hardware. Only in fiction will fossil dinosaur DNA, nurtured by a crocodile egg, bring forth a live dinosaur (or so we fervently hope); and sending aliens the genome of a cat is no substitute for sending the cat itself—complete with mice. An organism is, in fact, a self-organizing entity and more than the sum of its molecular parts. The informational metaphor all but ignores the multiple webs of relationships that make up physiology, development, evolution and ecology. Still, the tape and its reader set in order a portion of what we know, and the power of the analogy is reinforced every time a gene is altered, knocked out or replaced in the quest for knowledge or profit; within limits, the metaphor works and instructs.

All the metaphors in common use have merit, but none is altogether satisfying; and as Kant already noted, none of them makes comprehensible the existence of even a single blade of grass. It must be significant that we still have no language that makes organisms look at home in the physical universe; they are evidently much queerer than we suppose. It may be the case that our understanding of physics and chemistry lacks an essential girder that, when found, will span the gap. But it is also conceivable that there is a deeper flaw in the contemporary research program—Schrödinger's program, if you will—of bringing the science of life wholly under the umbrella of the physical sciences.

COMPLEX SYSTEMS

Complexity studies is a fresh label for a well-worn pigeonhole: general systems theory, that was pioneered by Ludwig von Bertalanffy in the thirties, and searches for laws common to systems of all sorts, whether living or not. What is new is, of course, the advent of the computer, which rejuvenated this plodding subject and made it, to some degree, an experimental one (6).

Most of us understand intuitively what is meant by a "system": an entity consisting of interacting parts that are arranged in some definite relationship to each other. A bicycle, the planets in orbit around the sun, and Colorado State University are all systems; a lump of granite or a sandpile is not. Complexity is much harder to grasp, and one of the perennial topics in the literature is just what makes a system complex, as distinct from merely complicated. Formalities aside, complexity is not hard to recognize and is commonly more a matter of degree than of kind. Diagnostic features include the emergence in the system as a whole of properties that cannot be assigned to any one of its components, invariance of the whole even though its components fluctuate, and a complementary interplay between local causes and global ones, such that each level constrains the other. Complex systems are commonly (though not necessarily) dynamic rather than static, and open to the input of energy and matter from the environment. Above all, they always display "some kind of non-reducibility: the behavior we are interested in evaporates when we try to reduce the system to a simpler, better-understood one" (7). Living systems, of course, represent the epitome of complexity by these (or any other) criteria, but they are not alone: a flame, a whirlpool, many electrical circuits and the circulation of currents in the oceans and the atmosphere come to mind. Even a sandpile can display complex behavior.

Robert Rosen (8), in what is perhaps the most rigorous and radical critique of the mechanistic approach to biology, has spelled out what the irreducibility of complex systems consists of. First, a complex system cannot be fractionated: there is no one-to-one relationship of parts to functions because one or more of the parts play several roles at once.

Second, while aspects of a complex system may have simple mechanistic descriptions, there exists no such description that embraces the system as a whole. Third, even those apparently simple partial functions change over time and diverge from what would have been their behavior in isolation. For all those reasons, complex systems are in principle not wholly reducible to simpler ones, and the Newtonian paradigm cannot be applied to them. By way of a pertinent example, biochemists may find it instructive to reflect on the folding of a linear chain of amino acids into the three-dimensional shape of the corresponding protein. Most proteins know how to fold up correctly, quickly, and spontaneously, but determined efforts to predict the final form by summing up the interactions of individual amino acids have been largely unproductive. Could it be, as Rosen believes, that the fault is not in the calculations but in the way the problem has been perceived? It is intriguing to read of novel and more holistic approaches (9), based on the premise that the amino acid chain is a complex system engaged in the molecular equivalent of morphogenesis.

A candle and a university can both be regarded as complex systems, but neither is alive; what then defines the subclass of complex systems to which organisms belong? Rosen seeks criteria that will be universally applicable to any form of life, even to life beyond the solar system or to fabricated organisms. Such criteria will be independent of any particular material incarnation, and must be drawn from those abstract principles of organization that make living systems living.

For Rosen, the heart of biology is that it revolves around the pattern of connections between components, and that allows him to offer a solution to Schrödinger's riddle: "A material system is an organism if, and only if, it is closed to efficient causation" (8). That is to say, if *f* is any component of a living system and we ask what is the cause of *f,* the question has an answer within the system. This would obviously not be true of the bicycle, and only partially true of the candle or the university. Actual organisms will be realizations of this general and abstract principle of organization. Note that Rosen, pursuing his quarry by formal logic, arrives at an insight remarkably like that attained by Maturana and Varela before him (Chapter 2): living organisms are autopoietic systems, they make themselves. Incidentally, in Rosen's view evolution is secondary: one can imagine life forms that did not evolve (e.g., fabricated ones), but evolution without life is inconceivable.

In an interesting and altogether constructive sense, Rosen can perhaps be described as a latter-day vitalist. His quest for the principles that make organic systems different form inorganic ones does not lead him to invoke mysterious forces that breathe life into the common clay, but he does bid us to rethink the relationship between biology and physics, and that is quite radical enough. Both disciplines deal with systems, and for the past two centuries biologists have sought to interpret their subject by the extension of laws inferred by physicists from the study of simple mechanisms. That, in Rosen's view, puts the cart before the horse: in reality, simple systems such as gases or planetary orbits are special and limited instances, while complex systems represent the general case. If organisms are ever to be understood as material physical entities, physics will first have to be transformed into a science of complex systems. This metamorphosis is already under way, but has proven neither quick nor painless: after half a century, the thermodynamics of irreversible processes (those that predominate in the real world) has chalked up few concrete achievements and remains largely outside the main stream of both physics and biology. I am not at all certain where this line of inquiry can lead; but Rosen's viewpoint will intrigue anyone who suspects, as I do, that the elusive relationship between physics and biology holds the key to Schrödinger's riddle.

*   *   *

Granted, then, that living organisms make up a subclass of the category of material, complex, dynamic systems; does this assignment entail any specific expectations concerning their properties, behavior, or genesis? We should not look to formal logic to predict particulars of form and function, which are bound to have a large element of the contingent, but we can hope for general, statistical features such as those that make all river valleys alike even though each one is unique. This project has already generated a large and confusing literature, and a growing number of prophesies. Of these, the one I find most convincing is also the most general: complex systems of the proper sort generate order spontaneously. The nature of this order varies. It may be spatial, as in the case of Prigogine's dissipative systems; examples include the convection cells that arise in a heated pan of oil, and the traveling waves displayed by certain chemical reactions (Chapter 7). But order can also refer to the interactions among elements of a network, such as those studied by Stuart Kauffman. The implication is that natural selection is not the sole source of order in the living world, but complements order that arises by the self-organization of complex systems.

   Kauffman's books (10), written in a high-colored and at times oracular style, display both the power and the limitations of a computer-driven approach to reality by way of the utmost abstraction. Imagine a huge network of interacting elements, 100,000 of them, these may be pixels on a screen or light bulbs. Each element is linked to others, perhaps just to one or two others or to dozens of them, and its response is governed by the rules of algebraic logic. For example, a particular element may flash when both of its two inputs turn on (a logical function corresponding to AND); another element may respond whenever either one of its two inputs turns on (an OR gate). Both the pattern of linkage and the rules that control any one element are assigned at random. We now start the network in some arbitrary state, and ask whether any regular pattern can be made out in the twinkling of lights. A naïve observer may well expect sheer chaos, lights popping on and off at random like a "berserk Christmas tree," but this is not necessarily what happens. The network's behavior depends strongly on the coupling rules, first of all upon the average number of inputs to each element. When that number is large, the network does indeed fall into apparent chaos. Conversely, when each element receives input from only one other, a simple fixed sequence quickly emerges. But when the number is a little larger, around two, something unexpected happens: the system settles down into a cycle consisting of a small number of states, about three hundred of them, which then repeats indefinitely. The cycle is robust: flip one element into its alternate state, and the network soon returns to its previous pattern. It displays something akin to homeostasis. Kauffman thinks of such cycles as "attractors" in the space of possible configurations that the network can adopt, and notes that they represent a minute fraction of an astronomically large number. Such compression represents "vast, vast order" arising of its own accord, guided neither by selection nor by intelligence.

Does this sort of abstruse model-mongering have anything to do with the birds and the bees? On the face of it, no—but wait. The genetic circuitry of a cell can be seen as a network of logical elements that switch each other on and off, and Kauffman claims that such networks will spontaneously adopt highly ordered regimes, even in the absence of natural selection. Spontaneous self-organization, "order for free," then becomes a background of regularity upon which natural selection acts, and against which its effects should be measured. Indeed, self-organized order will persist even in the face of pressure from natural selection.

The order that arises from self-organization, and that laboriously built by variation and selection, should not be seen as mutually antagonistic. On the contrary, Kauffman and others argue that natural selection can only succeed with logical networks that lend themselves to adaptive change by small incremental steps. Michael Conrad put it thus: "Why does evolution work? The reason is not to be found solely in the magic optimizing power of natural selection. It is as much due to the organizational structure that undergoes the variation. Evolution works because this organization is amenable to evolution, and because this amenability itself increases in the course of evolution" (11). Evolving networks shift gradually into modes of operation that are neither rigidly frozen nor chaotically fluid, but lie in the zone of transition between these regimes: evolution succeeds at the "edge of chaos." If all this is true (and proponents do emphasize that they are, themselves, walking the edge of uncertainty), real insights into the nature of living systems are emerging from the mist.

There is an evangelical earnestness to much of this literature that cannot fail to lift the skeptic's eyebrows. Earthlier biologists, our heads heavy with observations and experiments, are entitled to ask whether these theoretical generalizations make concrete and testable predictions. Indeed, there are some—albeit not a large crop. The number of states that a network can adopt is a function of the square root of the number of elements; could this be a deep law that underlies the rough correlation between the number of cell types in an organism and the square root of the number of genes (three for yeast with 6,000 genes, 300 for humans with 100,000)? It is also generally the case that the number of regulatory inputs per gene is small—though students of eukaryotic transcription can cite numerous exceptions. And there are some general features of evolution that would be expected from the theory of dynamic networks. Depending on whom one reads (12), these include the emergence of self-replicating and self-maintaining entities; developmental constraints on evolutionary change; punctuated equilibrium; the evolutionary laws of Dollo and Von Baer; even an innate tendency for complexity to increase with the passage of time. But it is only prudent to heed the warning flags before stepping into the deep water. Is it really true that a living system exhibits only a small number of stable states among which selection must choose? If so, laws of order do indeed constrain what natural selection can accomplish, but the astonishing diversity of living forms testifies to an all but untrammelled liberty. And is it really the case that the high degree of biological order could never have been achieved by random variation and selection alone? The life history of, say, the sea urchin displays the kind of capricious adaptations that natural selection would cobble together but highlights no global law of order.

On the outer banks of science, one often suspects that the believer is happy while the doubter is wise; and yet, too critical a spirit is apt to overlook the genuine contribution that complexity studies have already made. The great virtue of systems-thinking is not that it predicts the facts of life, but that it blurs that crisp line which divides the organic world from the inorganic. There is nothing mystical or unnatural about complexity, self-organization, emergence and wholes that are greater than the sum of their parts. These are properties of a large and diverse category of physical systems; even sandpiles do it. Organisms remain special, of course, thanks to their autopoietic character. But when organisms are seen as complex dynamic systems of a peculiar sort, the difference between the organic and the inorganic seems just a little less daunting. One feels encouraged to wonder just how autopoietic entities might have emerged from the much larger category of complex dynamic systems, and here energetics holds the most promising clues.

LIVING BY THE SECOND LAW

Thermodynamics and evolution, the two branches of science that revolve around order and time, started out in opposite directions: one views the world as running down, the other as building up, and we are still striving to close the circle.

First came that unfortunate collision over the age of the earth. William Thompson, Lord Kelvin, the first to recognize the central place of energy in physics, used the rate of heat production by the earth to calculate that it could not possibly be much more than 20 million years old; Kelvin allowed that his calculation may be in error if some internal source of heat existed, but none was apparent. Darwin never doubted that the earth was much older than that, but the dismissal of the very possibility of evolution for lack of time by one of the century's most eminent scientists cast a blight over his later years. Not until the end of the century was the issue laid to rest with the discovery of radioactivity, which granted the earth an age in the billions of years.

A more persistent conflict stems form the prime characteristic of living organisms: their ability to grow, develop and evolve, generating mounting levels of order in apparent defiance of the second law of thermodynamics. That most basic of natural laws mandates that all real processes be accompanied by the degradation of energy and the dissipation of order, and some creationists still insist that, therefore, thermodynamic principles preclude Darwinian evolution. Their objection stems from a misreading of the second law, which was corrected by Erwin Schrödinger (among others): Organisms extract "negentropy" from their surroundings. More precisely, organisms draw upon external sources of free energy to support their various work functions, including evolution. Even though the potential energy of the universe may be running down (and the continued expansion of the universe calls even this hoary verity into question), there is no reason why order cannot increase in favored locations. One such is the earth, bathed in a stream of energy from the sun. Just as the law of gravity does not forbid the spray to rise from the foot of a waterfall, the second law does not forbid the local generation of molecular and organismic order; it only requires that a local increase in order be properly paid for by the production of more disorder elsewhere. All the same, even with honest energetic book-keeping, there remains a troublesome discrepancy: classical thermodynamics would never lead one to *expect* a world that contains bugs, rainforests and all that lies between. It is this unconformity that is being smoothed out thanks to the advent of "irreversible thermodynamics," the science of open systems. We can begin to see that, far from conflicting with biology, the spontaneous tendency of energy to disperse and lose coherence as mandated by the second law may be the ultimate cause of all the complexity, diversity and beauty of the organic world (13).

Let me restate this point, for the matter is more subtle than it appears and quite fundamental. No one doubts that a supply of energy is necessary for organisms to persist, multiply and evolve. The molecular and physiological devices by which energy is harvested and coupled to the performance of useful work make up the discipline called bioenergetics (13); without mechanisms of this kind, life could not exist. The open question is whether the role of energy is merely permissive or actively causal: is energy required only for the workings of life, or is it the driving force for the emergence of complex autopoietic systems and their subsequent evolution? Those who envisage a fundamental link between the thermodynamic arrow of energy dissipation and the biological arrow of the greening earth make up a small minority, and stand well outside the main stream of contemporary biological science. But if their vision is true, it reveals that deep continuity between physics and biology, the ultimate wellspring of life.

The argument has been fully laid out by Jeffrey Wicken in a seminal but exceedingly dense book, and in a more accessible form by Bruce Weber, David Depew and Stephan Berry (14). Their central claim is that the emergence of complex systems and of life on the primordial earth can only be understood as a result of thermodynamic drive. We have already noted the appearance of physical dissipative structures in systems kept far from equilibrium by a continuous input of energy (convection cells, hurricanes, flames). The key to the emergence of life is the generation of structure and dynamic patterns in the chemical realm. The stream of energy, whether geochemical or solar in origin, progressively charged the primordial earth with chemical potential. It promoted first the formation of basic organic reagents (formaldehyde, hydrocyanic aid), and later that of more elaborate molecules such as amino acids, purines and even proteinoids (protein-like polymers of amino acids; Chapter XI). Energy flow also forced the appearance of larger, membrane-bounded entities, some of which became the precursors of protocells. In this view, rudimentary forms of metabolism and of cellular structure preceded the invention of informational macromolecules; but it is only with the advent of the latter that one can speak of Darwinian evolution and of life as we know it.

It may not be immediately obvious just how a stream of energy in the form of heat and radiation, impinging upon an inorganic earth of the proper chemical makeup, can (nay, must!) generate higher forms of molecular and structural organization. Setting aside the particulars of chemistry that account for the formation of this molecule or that, the reason that a flow of energy through a system organizes that system is that the buildup of complexity promotes entropy production and energy dissipation. The more efficient a given system, the greater the share of energy flux and chemical resources that it can command. Reaction cycles are a case in point. Consider a particular molecule, X, that absorbs light, and is thereby converted to another molecule, Y, with concurrent degradation of the light energy to heat. No such process can long continue unless a pathway exists that regenerates X from Y. Systems that "discover" such a cycle can persist, those that do not will soon cease to operate; thermodynamics favors the cycle even though it may entail several linked steps, so long as the cycle as a whole dissipates energy. Structural complexity, likewise, may be energetically favored. Biochemistry knows many instances in which small elements associate spontaneously into a larger complex; this entails a local increase in order but proceeds without any input of energy if the aggregation results in an increase in entropy (disorder) for the system as a whole. A case in point is the spontaneous formation of spherical lipid bilayer vesicles from free phospholipid molecules; the entropy increase is due to the liberation of water molecules. The devil, as usual, is in the details but the general principle holds: energy dissipation mandated by the second law supplies a driving force for the local generation of chemical and physical complexity.

Let me underscore a point that applies to all thermodynamic arguments. The existence of a driving force means that the process in question *can* happen, but does not ensure that it *will* happen. Energetics makes no predictions concerning the actual course of events. Energetics speaks to what is possible, actualities are a matter of kinetics and therefore of mechanism. Energy does not determine the particulars, but it can supply an ultimate cause for the emergence of structure on a lifeless earth. "Whereas the universe is steadily running downhill in the sense of depleting thermodynamic potential, it is also running uphill in the sense of building structure. The two are coupled through the second law" (15).

With the general course now set, Wicken goes on to reflect on the inexorable workings of the second law as the background to the proliferation and diversification of life. What are mutations, chromosome rearrangements, and other mishaps that disrupt the precise propagation of living order, but consequences of that randomizing tendency mandated by the second law? Not only variation, but natural selection as well, manifests the second law according to a general principle first articulated by A. J. Lotka eighty years ago. Natural selection does not evaluate organisms in pure culture, but operates on communities or ecosystems composed of many kinds of organisms evolving together. Every ecosystem can be regarded as an energy transformer, trapping radiant energy in the form of biomass and gradually degrading it to heat. As this transformation proceeds, the ecosystem makes available niches within which particular organisms can flourish and multiply according to their capacity to acquire a share of that energy stream. Wicken, like Lotka before him, holds that natural selection favors those organisms that are most effective in channeling the flux of energy through themselves and, concurrently, in increasing the flux of energy through the ecosystem as a whole. Selection, in this view, operates hierarchically on many levels: not only on individual organisms struggling selfishly for immediate advantage, but also on populations (species) and the communities in which they participate. Thanks to its roots in thermodynamics, the general course of evolution becomes predictable: there is a tendency for energy flow (and for life) to expand into any niche, provided there is a mechanistic path; it will diversify, radiate, speciate; and it will tend to produce structures that are increasingly complex. In Wicken's view, evolution need not struggle relentlessly against the forces of decay but goes with the natural flow of the universe. The manifold faces of life appear as features to be expected, not as implausible marvels that must be explained. But we must be careful not to confuse the general with the particular: biological details cannot be deduced from the second law, since organisms riding the energy stream still remain creatures of history.

It will be obvious that this view of the logic of evolution is substantially at variance with neo-Darwinian orthodoxy. The proposition, stated or implied by Lotka and some other proponents of the thermodynamic vision (though not by Wicken), that natural selection directly rewards organisms that maximize energy flow rather than reproductive fitness, sticks in many a craw including mine. But the wholesale rejection of this entire line of thought as being teleological, and therefore unscientific, strikes me as excessively dogmatic. In particular, the assertion that selection acts hierarchically on levels both lower and higher than the classical individuals (on genes, for instance, and on species), has too many reputable supporters to be dismissed out of hand. I am impressed by Wicken's assertion that "Individuals are the proximate carriers of structural information. But that information includes the survival strategy of a *species*, whose ecologically contexted history is impressed into adaptive strategies of gene carriers" (15). In that sweeping sense, it is legitimate to argue that "The most general units of selection are not individuals, but informed patterns of thermodynamic flow, of which organisms, populations and ecosystems are all exemplifications" (15).

What we have here, as we often do on the marches of science, is not a hypothesis that can be falsified by experiment but a heuristic viewpoint. The thermodynamic perspective reveals some features neglected by the genetic one, while obscuring others. What is needed to reconcile the different viewpoints are explicit ideas of how the struggle among selfish individuals for reproductive success shapes the patterns of energy flow in a mixed biofilm, or in a summer meadow glad with flowers and bumblebees. Such efforts are indeed underway (16), but lie outside the bounds I have set for this book.

Does the scientific spirit compel one to proclaim one viewpoint true and the other false? I do not think so, for one addresses what is possible and the other sees only the actual. For an everyday parallel, reflect for a moment on that hero of popular American culture, the self-made man. It would be churlish to deny him the virtues and rewards of intelligence, hard work and good fortune. But common sense demands that we acknowledge his circumstances, including a society that cherishes free enterprise, political liberty and the rule of law. Would our hero have been equally successful had he been born into a peasant hut in an Indian village?

A UNICORN ON THE WALL

For thirty years and more, the mascot of my laboratory was a unicorn resting placidly in a fenced enclosure, a reproduction of one of the magnificent tapestries held by the Metropolitan Museum of Art in New York. It provoked all sorts of ribald comments, but for me the symbolism was never in doubt. The unicorn stood for the unknown in nature, that ineffable strangeness that scientists strive to corral and tame. Schrödinger's riddle is central to that quest.

To the question, What is Life?, science presently offers two answers. The first asserts that living organisms are autopoietic systems: self-constructing, self-maintaining, energy transducing autocatalytic entities. The alternative answer proclaims that living organisms are systems capable of evolving by variation and natural selection: self-reproducing entities, whose forms and functions are adapted to their environment and reflect the composition and history of an ecosystem. The two answers are not identical, but there is substantial overlap between them; they emphasize different aspects of a rounder reality. One can imagine autopoietic entities that did not arise by evolution (the astronauts in Arthur Clark's space epic 2010 learn of some, seeded on Jupiter for the purpose of turning it into a lesser sun), but we would probably label them robots rather than organisms like ourselves. By the same token, RNA molecules in a test-tube copied repeatedly with the aid of a replicase enzyme clearly evolve by variation and selection, but would hardly be considered alive; even viruses represent a borderline case. The best answer we can offer at present combines both partial ones (17): life is the property of autopoietic systems capable of evolving by variation and natural selection. This definition will do for the present, but may need revision as our technology creates more and better Golems, or when we gain experience of universal biology.

Some years back, ears pricked up around the world when scientists from the National Aeronautics and Space Agency announced the discovery of putative microfossils in meteoritic rocks thought to have come form Mars. That publicity release seems to have been erroneous, but now comes news of planets encircling several stars in the Milky Way, the first beyond our own solar parish. Astronomers, and indeed most scientists including myself, take it for granted that planets are common in our galaxy and that some, at least, will harbor life. In company with the majority of science fiction writers, we reject the claim that life must be the product of some desperately improbable constellation of events that did occur once but can never come again—a miracle, in effect. On the contrary, many of us hold that life is the probable outcome of physical and chemical processes that are likely to take place on any planet of the proper size, location and composition. In other words, we take for granted that emergence of biology from physics and chemistry whose lineaments still elude us (Chapter 11). This is a leap of faith that may prove to be misplaced; in the meanwhile, it provides a focus for reflection on what it means to be alive.

The question is tantamount to asking whether there are universal laws of biology, over and above those of physics and chemistry; or at least, qualities that we can expect to see in any form of life. In my opinion, not universally shared (18), the answer must be yes: the very definition of living things as autopoietic systems capable of evolution implies quite a number of general features of organization. Life everywhere is likely to be a property of organisms: material, dynamic patterns in space and time that exist by virtue of a stream of matter, energy and information. To live means to eat, metabolize, reproduce and pass away; birth and death are part of the natural order. Another candidate for universality is cells, or something like them: units of reproduction and selection, separated by a semipermeable boundary from the environment and from each other. In heaven as on earth, I suspect, like begets like and organisms come only from pre-existent organisms. And life everywhere is bound to be complex. The mathematician John von Neumann famously calculated what would be required to make a self-reproducing entity *in silico*. The screen would have to hold some 10^{13} pixels; assuming 1 mm² for each pixel, it would be about 3 km across (19). No wonder that program has never yet been run.

That innocuous phrase "capable of evolution," is likewise freighted with implications, including genesis from an inorganic world by the agency of thermodynamic drive. Once an autocatalytic system arose it will have multiplied explosively, preempting later rivals. One can expect that life, wherever it is found, will take the form of a comprehensive web of organisms that are linked both by descent and by their metabolic economies. They will display diversity and the effects of contingency, but will be effectively adapted to whatever niches their environment supplies. Moreover, life will be made up of individuals of sorts: units that function as coherent wholes, are subject to variation and provide the normal target of natural selection. From this Olympian viewpoint it looks a lot like the world we know and not nearly as exotic as a medieval bestiary! Some day we may find out whether the foregoing represents valid inferences form the study of life on earth, or merely betrays the poverty of one writer's imagination.

So much for organization; what commonalities can we expect on the material plane? Terrestrial biochemistry revolves chiefly around combinations of a small stock of elements: carbon, hydrogen, nitrogen, oxygen, phosphorus and sulfur (CHNOPS), plus a set of inorganic ions.

The reason is that these elements (and particularly carbon) readily form bonds with themselves and each other, and can thus support a vast range of molecular forms and activities. The higher elements are much less sociable; there has been some speculation about alternative biochemistries (based on silicon, perhaps) but that seems far-fetched. CHNOPS, I expect, will prove universal—and so will water. That argument suggests, in turn, that we may be able to recognize a living planet by the peculiar composition of its atmosphere (20). The major pathway of terrestrial photosynthesis makes use of water as reductant and generates oxygen as a byproduct. It seems quite likely that the same principle has also been discovered elsewhere, perhaps with organisms that utilize oxygen for respiration as additional byproducts.

Beyond this point the argument becomes increasingly tenuous. Molecules that are readily formed by planetary chemistry (including glycine, alanine and purines) are likely to play some role in the emergence of life, but more intricate structures may be unique to this locale or another. What about chlorophyll, RNA, DNA, glucose, ATP, and proteins made up of the standard set of twenty amino acids—are these universal building blocks or merely terrestrial? The answer depends on whether these structures are probable products of a prebiotic chemistry that has not yet been discovered, or conserve some kind of historical accident. I incline toward the latter view, but suspect that commonalities may exist at a more abstract level: linear genophores, catalysts with corrugated surfaces and fluid membranes. For universal biology, shape and physical parameters may be more significant than chemical makeup. This speculation illustrates, at the molecular level, what Daniel Dennett refers to as "forced moves in design space": structures that are likely to be discovered by evolution again and again, because they do the job.

Could it be that the same argument holds for legs to walk on (rather than, say, wheels or tank-treads), wings to fly with, eyes in the front of a head and brains to keep it all coordinated? These musings provide some corrective to the view so powerfully stated by S. J. Gould on many occasions (21) that the life we know is unique and will never be recreated elsewhere; intelligent life, in particular, is a fluke that will be vanishingly rare in the universe. Time, perhaps, will tell whether contingency rules or bows before the power of natural selection to create successful design. As for myself, I hope for an alternative biosphere in which a unicorn in the garden is just what one would expect.

11

SEARCHING FOR THE BEGINNING

"And the end of all our exploring
Will be to arrive where we started
And know the place for the first time."

T. S. Eliot, *Little Gidding*

FRAMING THE PROBLEM
STARTING WITH THE SOUP
MOLECULAR PREHISTORY AND THE RNA WORLD
HOT SPRINGS AND FOOL'S GOLD
A BRIDGE TOO FAR?

Of all the unsolved mysteries remaining in science, the most consequential may be the origin of life. This opinion is bound to strike many readers as overblown, to put it mildly. Should we not rank the Big Bang, life in the cosmos, and the nature of consciousness on at least an equal plane? My reason for placing the origin of life at the top of the agenda is that resolution of this question is required in order to anchor living organisms securely in the physical world of matter and energy, and thus relieve lingering anxiety as to whether we have read nature's book correctly. Creation myths lie at the heart of all human cultures, and science is no exception; until we know where we come from, we do not know who we are.

The origin of life is also a stubborn problem, with no solution in sight. There is, indeed, a large and growing literature of books and articles devoted to this subject, many with theories to propound (1). Biology textbooks often include a chapter on how life may have arisen from non-life, and while responsible authors do not fail to underscore the difficulties and uncertainties, readers still come away with the impression that the answer is almost within our grasp. My own reading is considerably more reserved. I suspect that the upbeat tone owes less to the advance of science than to the resurgence of primitive religiosity all around the globe, and particularly in the West. Scientists feel vulnerable to the onslaught of believers' certitudes, and so we proclaim our own. In reality, we may not be much closer to understanding genesis than A. I. Oparin and J. B. S. Haldane were in the 1930s; and in the long run, science would be better off if we said so. After all, the unique claim of science is not that it has all the answers but that it knows the questions, and will not compromise its commitment to the rational search for truth.

What makes the origin of life quite so intractable? The object is to discover what transpired in the exceedingly remote past, under circumstances that one can hardly imagine. The clues to be read in the rocks and in the constitution of contemporary organisms are tenuous at best. Experimental approaches are inviting, but handicapped in principle: they can suggest possibilities but cannot reveal the actual course of events. The ultimate stumbling block may be the conceptual one. Life is, by definition, a quality of complex systems that generate themselves and their constituents and undergo evolution by variation and selection. What we seek to understand is not the origin of particular molecules, but of the kind of functional organization that could multiply, diversify, and inherit the earth. It is a very hard nut to crack.

By way of compensation, this is truly a frontier that calls to adventurous spirits. Societies of molecular and cell biologists number their members in the tens of thousands, while the International Society for the Study of the Origin of Life claims four hundred; needless to say, funding for such blue-sky research is hard to come by. But this is science as it was in the old days. The goal is not to fill in the corners but to discover one of the deepest secrets of the universe, and therefore reflection has not yet been swamped by the flood of information. There is no guarantee that the quest will be successful, and little expectation of material rewards, but the origin of life remains a grail worth seeking.

FRAMING THE PROBLEM

Life arose here on earth from inanimate matter, by some kind of evolutionary process, about four billion years ago. This is not a statement of demonstrable fact, but an assumption almost universally shared by specialists as well as scientists in general. It is not supported by any direct evidence, nor is it likely to be, but it is consistent with what evidence we do have. The contrary claims of Biblical fundamentalists can, for the most part, be clearly refuted on the geological evidence. All the same, it is important to acknowledge the degree to which this field of inquiry is founded on surmise. The reasons for the general consensus are, first, the lack of a more palatable alternative; and second, that absent the presumption of a terrestrial and natural genesis there would be no basis for scientific inquiry into the origin of life.

Consensus on the principle, while more solid than it was a few decades ago, leaves plenty of room for disagreement. One bone of contention is the postulate that life arose on earth. Ever since the Swedish chemist Svante Arrhenius proposed at the beginning of the twentieth century that life is universal in the cosmos and that its seeds arrived here on the backs of meteorites or dust particles ("panspermia"), some have sought the source of life far beyond earth. It is now considered unlikely, but perhaps not altogether impossible, that bacterial spores could have survived a space journey embedded inside a rocky matrix. By way of a radical alternative to the conventional wisdom, Francis Crick has made the case for "directed panspermia," the proposal that seeds of life were dispatched here intentionally by a galactic civilization (2). I don't suppose that Crick himself ever believed in this notion; his purpose was rather to demonstrate that the arguments for a deliberate sowing of life are no weaker (and no stronger) than those for a terrestrial origin. The story's latest twist calls for a shipment from Mars (3). Meteorites found in Antarctica, and thought to be of Martian origin, contain objects that may conceivably represent fossil microorganisms. Mars is smaller than earth and has a weaker gravitational field. If life evolved there, organisms (or more likely, spores) sheltering within Martian rocks might have been ejected by a meteorite impact and survived the short hop to earth (the converse route would have been far less likely). It would certainly be wondrously exciting if life, past or present, were discovered on Mars; but the notion that terrestrial life originated there seems too far-fetched to be worth pursuing here.

The timing of life's inception, while uncertain, is definitely constrained by geological data. There is good reason to believe that the earth formed about 4.5 billion years ago. Whether the infant earth was torridly hot or blanketed with ice is still under debate, but it was evidently subjected to heavy meteorite bombardment early on. Some of these impacts were powerful enough to set the oceans aboil (ejection of the moon is blamed on an earlier episode of this kind, involving collision of the earth with an object the size of Mars); even if life had already begun to evolve, it would presumably have been extinguished. The heavy barrage ended about 4 billion years ago, and the earth has been continuously hospitable to life ever since (setting aside the occasional "lesser" impacts, such as that blamed for the abrupt demise of the dinosaurs). On the near side of the window, the earliest unambiguous fossils (bacterial cells and stromatolites) have been dated to 3.5 billion years ago (some would prefer 3.2 billion). Tentative traces of still more ancient life come from the Isua formation of Greenland, thought to be around 3.8 billion years old. These rocks have been subjected to heating and compression and lack identifiable fossils, but they do contain organic material with a relatively high C^{12}/C^{13} ratio. Such isotopic enrichment is also seen in living matter, suggesting that the organics of Isua may be of biological provenance. Taken together, these findings imply that life originated during a relatively short span of no more than 500 million years, about as much time as separates us from the explosion of animal life at the onset of the Cambrian era. Whether that is plenty of time or a little tight is hard to judge: until the process is understood in principle, one cannot say whether genesis should be measured in aeons or in decades.

Geology is rather less helpful with the conditions that prevailed on the young earth; that information is critical, for the physical setting determines what makes a plausible prebiotic chemistry. Biologists have long nursed the hunch that life arose in the ocean, but it is not at all clear when the ocean formed, how hot it was, and whether it was salt or fresh. Darwin himself, in a personal letter to a friend, mused wistfully on a "warm little pond," but it now seems unlikely that such a benign place could have existed 4 billion years ago. The state of the atmosphere has been the subject of special attention. All agree that oxygen, which makes up a fifth of today's atmosphere, was produced by photosynthetic organisms over the past 2 billion years; no more than traces of oxygen can have been present when life arose. What about the other gases? The pioneers of planetary history favored a strongly reducing atmosphere containing methane, ammonia, and hydrogen gas in addition to nitrogen and CO_2. Such an atmosphere would encourage the formation of reduced organic substances, including some of biological relevance, and it underpins the widespread belief that the primordial ocean consisted of a dilute broth of ready-made biological precursors. By contrast, contemporary work suggests a nearly neutral atmosphere produced by volcanic activity and consisting primarily of CO_2, nitrogen, and water vapor. Organic substances do not readily form under these conditions; the new findings call into question the "myth of the primordial soup" (as Robert Shapiro puts it), and undermine all scenarios of genesis that call for a long period of chemical evolution to generate the building blocks of life (4).

The geological setting would not matter so much if the origin of life were ultimately a chance event, frightfully rare perhaps but one that did happen on a singular occasion—and here we are to marvel at our good fortune! The probability of an event is, after all, a function of both the likelihood of each toss, and the number of trials. Given billions of years and untold trillions of molecules to collide, separate, and meet again, would it not be conceivable for a protocell of some kind to arise by chance (or, to skew the odds a little, a self-replicating molecule to set the ball rolling)? It is, in principle, impossible to rule out this fantasy, which lurks at the base of some popular notions about how life may have begun. But a chance origin commands much less respect than it did a decade or two ago, for two reasons. One is that both of these events are so enormously improbable that the entire universe had not time and atoms enough to conduct all the trials called for by the hypothesis (5). The other reason for rejecting a chance origin out of hand is that the postulate would terminate the inquiry: science cannot really deal with unique events, which are effectively miraculous.

We are thus left with the presumption that life arose or evolved from inanimate matter by physical and chemical processes that are, in principle, comprehensible and discoverable—perhaps even reproducible. Just what these processes may have been is what the study of biopoiesis is about. It bears repeating that we know very little for certain, and that it is seldom possible to formulate hypotheses that can be falsified by experiment; the opinions of scholars are, therefore, colored by personal beliefs about what should have happened, and even about what is meant by "life." Chemists tend to hold that once the molecular building blocks were on hand most of the heavy lifting had been done, and therefore, seek enlightenment in prebiotic chemistry. Geneticists, preoccupied with nucleic acids, insist correctly that evolution requires heredity, variation, and selection; and they therefore define the problem in terms of the emergence of biological information. And physiologists are adamant that without compartments and a sustainable source of energy one cannot make nucleic acids or anything else. It is almost irresistible to see these three levels of complexity as consecutive stages on the road to the emergence of full-fledged cells, and that idea has provided a conceptual framework for much of contemporary research on the beginnings of life.

Starting with the Soup

Forty years ago, a spate of discoveries promised to bring an explication of genesis within reach, or at least to make this quest a province of experimental science. The prevailing conception at the time, formulated two decades before by A. I. Oparin, in the Soviet Union, and J. B. S. Haldane, in Britain, was that life arose spontaneously in an environment replete with all sorts of organic substances, such as are produced today only by living organisms. The young earth would have been a very different place from the one we know. Its atmosphere was not only devoid of oxygen but strongly reducing, and thus favored the reductive reactions required to produce, say, amino acids; there were no scavengers to gobble up organic substances as soon as they were formed; and the roiling globe supplied abundant energy in the form of heat and lightning. Organic matter would, thus, accumulate in the oceans or in lagoons and tidal pools, undergo further transformation, and over many millennia come to make up a rich broth containing all that was required to bring forth the first protocells.

In 1953, Stanley Miller, then a graduate student at the University of Chicago, set up a simple experiment to assess this hypothesis and determine what kinds of molecules might have been present on the early earth. His apparatus consisted of a closed set of interconnected vessels filled with a mixture of methane, ammonia, and hydrogen gas, corresponding to what was then thought to have been the composition of the primordial atmosphere. In one of the vessels, water was kept on the boil; the vapor was subjected to continuous electric discharges (to simulate lightning), condensed and returned to the reservoir. After a few days, the water turned brownish and tar began to deposit on the vessel walls. When the solution was analyzed it proved to contain an array of organic molecules. Glycine and alanine, amino acids prominent in proteins today, were most abundant; aspartic and glutamic acids were also present, together with traces of other amino acids and varying quantities of compounds not found in proteins, such as α-aminobutyric acid. A few years later Juan Oro' showed that analogous experiments under somewhat different conditions yielded not only amino acids, but also the purines adenine and guanine, constituents of all nucleic acids. Other investigators produced simple sugars, fatty acids, even polymers of amino acids akin to proteins. A major field of research, "prebiotic chemistry," was launched by those pioneering studies and quickly captured the imagination of scientists fascinated by the abiding mystery of biological origins (6).

Expectations soared with the discovery that certain meteorites, designated "carbonaceous," contain small amounts of organic matter, similar in composition to the products of prebiotic syntheses and definitely
of extraterrestrial origin. Optimists looked forward to the day when
most cell constituents would find plausible precursors in the ever-
thickening soup. A large and imaginative literature has grown up around
reactions that may have made up the primordial metabolism of the
earth, mediated not by enzymes, but by mineral catalysts such as clays.
Historically, most authors imagined this protometabolism to take place
within protocells, presumably generated by spontaneous assembly from
pre-formed components. The very first protocells would have lived by
the fermentation of substances abundant in their environment (glycine,
perhaps), and used other constituents to manufacture their own fabric.
As amino acids, purines, and fatty acids were depleted one by one,
metabolic pathways evolved backward from one available precursor to
the next. Eventually when fermentable substrates were exhausted too,
their role was filled by photosynthesis starting, perhaps, with abiotic
porphyrins from the soup.

Some contemporary authors go further, suggesting that protocells
need not be a prerequisite for the emergence of metabolism. In this
vein, Christian de Duve (7) has described in considerable detail how a
proto-metabolic web, akin to that of contemporary organisms, may have
arisen prior to the appearance of membranes and of cells. He calls upon
prebiotic peptides to serve as catalysts, and advocates a key role for
thioesters as precursors to the phosphorylated energy carriers of the present day. The oxidation of ferrous iron by illumination with ultraviolet
light may have been the ultimate source of reducing power; and proteins, nucleic acids and membranes were all products of the expanding
web of reactions rather than its foundation. Stuart Kauffman's proposal
(7), albeit quite different, begins from the same central premise. He
imagines a large pool of small peptides, generated by abiotic chemistry,
each capable of catalyzing some reaction (however weakly). If the number of such reactions becomes large enough, they will form a self-
organizing network that eventually becomes self-replicating. Kauffman's
hypothesis specifies neither enclosure nor even an explicit source of energy. What these two propositions, and the great majority of others,
have in common is their reliance on an abundant and varied supply of
prebiotic precursors to contemporary biochemistry.

Can it be true that molecular complexity came first, and provided
the foundation upon which cellular complexity arose by some kind of
self-assembly? The idea remains seductive because it promises to make

the problem of origins tractable by laboratory procedures, but the early confidence has been seriously dampened (8). The problem is not only that geologists have lost faith in the reductive atmosphere: meteorites carrying organic substances generated in outer space could have made good the shortfall in terrestrial production. A graver objection is that, while some of the simpler biomolecules form readily under conditions that can be loosely described as "prebiotic," most are only produced in trace amounts and others not at all. Purines are easy to come by, pyrimidines much less so; there is still no prebiotic route to cytosine, which is found in both DNA and RNA. Simple sugars are readily made from formaldehyde, but ribose is a minor constituent in a melange of sugars of all sorts; besides, nucleosides are hard to make. Fatty acids, too, have remained elusive. Proteinoids, polymers of amino acids linked by peptide bonds, form when a mixture of amino acids is heated to dryness, and they display weak but varied catalytic activities. But these polymers are branched rather than linear, and their sequences are not far from random. Phosphate, a major constituent of biological molecules that are involved in both heredity and metabolism, is also a major headache, for the presence of calcium or iron makes phosphate precipitate from solution. The only inorganic precursors for ATP and other phosphoryl donors are pyrophosphate and polyphosphate; they are rare in nature and do not ring true as candidate energy carriers.

Both promise and misgivings come to a head over the production, by credible prebiotic procedures, of a self-replicating informational molecule such as RNA. The hypothesis, that life began with naked protogenes that replicated themselves in the primordial broth and then "learned" to encode proteins that promoted the protogenes' multiplication, has been advocated with varying degrees of conviction and hesitancy by many of the leading lights in this field (9). It maintains almost a stranglehold on the minds of scientists and the general public alike, and fuels what is far and away the most intense research effort. The goal is to find conditions under which RNA will multiply in the test tube, replicating a particular sequence in the absence of enzyme proteins. The purposes that animate this enterprise go beyond the hypothesis that supplied its initial impetus: should the effort succeed, much will be learned about the chemical foundations of both replication and catalysis, and technological applications can be imagined (10). Few of the participants doubt that it can be done, and partial success has been attained. But no one has yet grasped the prize: to generate RNA from a mixture of activated nucleosides in the absence of enzymes, and have it supply the template for its own replication. Even if the next issue of

Nature heralds success, curmudgeonly physiologists will have questions to ask. Where, for instance, did those activated precursors come from on the primitive earth? What kept them and their products from diffusing away, or being degraded? Are the conditions for replication plausible in a geological setting? How can any self-replicating RNA molecule pick out the "correct" monomers from a broth that also contains chemically incorrect ones? Celebration will certainly be in order, but so will restraint and a soupçon of modesty.

Many of these misgivings would be allayed if the emergence of rudimentary metabolism, energetics, and heredity were conceived, not in free solution but in a compartment of some kind. In contemporary cells boundaries take the form of lipid bilayer membranes; and one can argue that lipids, too, were on hand in the prebiotic soup—perhaps not phospholipids, but short fatty acids (11). In a provocative series of experiments, David Deamer and his colleagues prepared lipid extracts from a carbonaceous meteorite; the lipids readily formed membrane vesicles, apparently composed of a mixture of short-chain fatty acids and polycyclic aromatic compounds. Vesicles spiked with fat-soluble dyes can be excited with light and then develop large differences in pH between the external medium and the lumen. One can write schemes that allow simple combinations of pigments and hydrogen carriers to translocate protons across the membrane; these in turn suggest possible stages in the evolution of chemiosmotic energy coupling (12).

Membrane vesicles, when dried and then rehydrated, will take up whatever substances were present in the mixture including proteinoids and nucleic acids. Attempts are presently underway to demonstrate nucleic acid synthesis inside vesicles, using precursors supplied from the outside. This is heady stuff, chemistry for gods. There is a caveat that applies to all proposals that interpose a membrane between the locus of action and the outside world: any membrane tight enough to sustain energy coupling by proton currents will also exclude ions, nutrients, and metabolites. Compartments demand specific transport carriers, proteins as a rule, whose provenance needs to be specified. But there may be a way out of the dilemma: perhaps the earliest protocells were inside-out ("obcells"), such that metabolic reactions and even the production of proteins and nucleic acids took place on the exterior side facing the primordial soup, while protons were translocated into the lumen. Fold the obcell upon itself, seal around the edges, and you have a cell of the proper polarity with two peripheral membranes, a possible precursor to the Gram-negative bacteria (12).

The hypothesis, that protocells somehow assembled themselves from a medium chock-full of prefabricated precursor molecules of the right sort, is now more than sixty years old; one or another of its many variants underlies virtually all of the experimental and theoretical literature produced in the field. Most of the attention has been devoted to the chemical tier, but the chasm that divides molecules from organized cells is wider still. Cells make themselves, but not by self-assembly of pre-formed molecules; they grow, thanks to a generative biochemistry that produces a restricted subset of molecules within a confined and structured space. If complex systems are to arise from raw chemistry, hold out against the forces of decay and multiply themselves in space and time, energy must supply the driving force. No biopoietic scheme deserves to be taken seriously unless it provides both an explicit and sustainable energy source, and plausible means to couple that energy to the flow of matter through the emerging system. What distinguishes the cell from the soup is the former's purposeful organization; how strange, then, to find the literature all but silent on the genesis of that organization!

Scientists like to measure the value of an idea by the volume of research that it has inspired, and by this criterion the prebiotic soup has done very well. But a historical theory must account for historical events, and in truth there is not (and perhaps cannot be) convincing evidence that there ever was a rich broth of organic substances, or that it played the role assigned to it by the theory. The hope now is that clever chemistry will point the way, if not to the origin of life, then to some kind of simulacrum thereof. That has not happened yet, and I am ambivalent as to whether one should be disappointed but hopeful, or just quietly thankful.

MOLECULAR PREHISTORY AND THE RNA WORLD

Prehistorians, with little more than scraps, shards, and analogy to go by, do not reconstruct the past so much as imagine a plausible version of it. That is doubly true of their molecular colleagues, for evolution's relentless drive for advantage seems to have erased virtually all traces of the origin of cellular life. In Chapter VIII we inferred the existence and characteristics of the last common ancestor(s) of all life from the general principle that ubiquitous features are bound to be ancient. Organisms at that stage of evolution, which must have been reached more than 3.5 billion years ago, were constructed along prokaryotic lines and had already acquired all the basic biochemical and physiological hallmarks of bacteria: genes inscribed in DNA, proteins for catalysis and work functions, ribosomes, cell walls and a mechanisms of division, lipid membranes studded with diverse transport systems, redox chains and ATPases, and chemiosmotic energy transduction. This conclusion seems quite solid, but we remain entirely in the dark concerning the origin of that basic molecular machinery itself.

Cell components as we know them are so thoroughly integrated that one can scarcely imagine how any one function could have arisen in the absence of the others. Genetic information can only be replicated and read out with the aid of enzyme proteins, which are themselves specified by those same genes. Energy is harnessed by means of enzymes, whose production requires energy input. Darwinian evolution is at bottom the struggle among individuals defined by cell membranes, yet how could membranes and transport catalysts arise without genes, proteins and energy? Here are profound mysteries, for unless one is willing to entertain a miraculous genesis, that universal ancestor of all cells must itself have been the product of a protracted and stepwise evolution, guided by variation and natural selection. The task of the molecular prehistorian is to seek traces of that passage to the only kind of life we know.

A sizeable literature already records attempts to call up a possible past from the arrangement of metabolic pathways and their genes, or from the divergence of transport ATPases; but these musings do not reach down to the ultimate question of how molecules came to life. For most biologists, the hallmark of life is the capacity for Darwinian evolution; and the most fundamental relationship is that which links nucleic acids, the carriers of information, to proteins that perform functions and generate the phenotype. If a single question can stand for the whole mystery, it must be the origin of this most intimate partnership.

Contemporary cells and organisms revolve around the trinity of DNA, RNA, and protein. But even thirty years ago, following the discovery that RNA serves as the genetic material in certain viruses and also participates in a number of physiological functions, molecular biologists began to suspect that life's earliest stages relied on a much simpler chemistry in which both replication and functions resided in RNA. This speculation was mightily reinforced in the eighties, with the discovery that certain steps in the processing of RNA transcripts are catalyzed, not by protein enzymes of the usual sort, but by RNA alone. This startling finding proved that RNAs could function as highly specific catalysts, and a string of such "ribozymes" has been discovered since. The hypothesis that, in the beginning, there was neither DNA nor protein, but only RNA that replicated itself and also catalyzed whatever functions were required, is referred to as the "RNA world." It enjoys wide currency among molecular biologists, and has become a cornerstone of biopoiesis research (13).

The arguments that support an RNA world of some kind are circumstantial, but they rest on a set of peculiar features that just may represent relics from a time when RNA reigned supreme. There is first the fact that, of all known macromolecules, RNA alone can direct its own replication (given the necessary enzymes) and also catalyze chemical reactions. RNA also participates in the replication of DNA, providing a primer to initiate copying and a ramp that lets it go to completion. Might these be vestiges of a machinery that once replicated, not DNA but RNA itself? That suggestion meshes with the long-known fact that the nucleotides from which cells make DNA are themselves generated by reduction of the corresponding RNA precursors, hinting that RNA synthesis is the more ancient process. Turning to catalysis, remember that RNA plays several crucial roles in protein production (Fig. 4.6), carrying information in some (messenger RNA, transfer RNA) but doing chemistry in others. It cannot be accidental that RNA, rather than a standard protein enzyme, performs the ribosome's essential task: linking the next amino acid to the growing peptide chain. There is also the curious fact that coenzymes, which play so many different roles in metabolism, are commonly modified ribonucleotides, ATP is a case in point (Fig. 4.4). It has been proposed that coenzymes may be relics of a time when all catalysis was the task of RNA; and to my knowledge, no one has made a more plausible suggestion.

Most proponents of the RNA world, Walter Gilbert in particular, envisage it as a pre-cellular stage of evolution. In this view, the first proto-organisms consisted of naked self-replicating RNA molecules that emerged somehow from the primordial broth. These molecules evolved, spontaneous mutants being selected for faster replication, and they "learned" to encode and construct proteins to assist in their replication. Eventually these free replicators became enclosed in membrane vesicles, and found ways to cooperate for the common good; cellular evolution could now begin. But an RNA world is equally plausible (or fanciful) as a stage in the evolution of cells. If we postulate that life was enclosed from the very start, the earliest kind of metabolism might be that required to support the production of nucleotides and the self-replication of RNA molecules; proteins came later, DNA later still. Both accounts quickly encounter difficulties. For all their efforts, chemists have been quite unable to put RNA together under "prebiotic" conditions, let alone make it replicate. It presently appears unlikely that these could have taken place by chance on the lifeless earth. Advocates of full ribo-organisms, all of whose activities would have been carried out by RNA, must acknowledge that the catalytic skills of RNA are more limited than those of proteins; RNA has little aptitude for the kinds of chemical reactions that underpin metabolism and energetics, not to mention membrane transport. Finally, regardless of the structural context, the rocky path from RNA replicators to DNA genes and from catalytic RNA to protein enzymes calls for stout boots and a good head for heights.

Was there really ever a time when ribo-organisms represented the seeds of a more versatile biology to come? I do not know. Here again, as in the case of the primordial soup, the fruitfulness of the hypothesis is not in doubt; what needs to be established is its historical veracity.

Hot Springs and Fool's Gold

One of the oddest findings of molecular phylogeny is that the deepest branches of the universal tree bear chiefly autotrophic and thermophilic organisms. Now this may turn out to be a temporary truth, or a red herring; but it could also convey real information about the world of the last common ancestor, and pass a broad hint about the circumstances in which life originated (14). The latter is also the opinion of Günter Wächtershäuser, a German chemist and patent attorney, whose iconoclastic hypotheses and philosophical broadsides have stirred up a field badly in need of fresh ideas. Where almost all investigators imagine a primordial broth of organic compounds from which sprang self-replicating macromolecules and then cells, Wächtershäuser insists that living systems were self-generating from the start: his is not a heterotrophic genesis but an autotrophic one.

Wächtershäuser has spelled out his proposals in a series of major papers and, most recently, in rebuttals to critical comments by others (15); the present summary is necessarily much condensed. He begins by rejecting out of hand the stock of prebiotic organic precursors: not only is its existence unlikely, its uncritical acceptance has encouraged bad scientific practice ("the broth became stronger and stronger, but the argument became weaker and weaker"). Instead, life arose directly from the mineral realm thanks to an original source of energy and reducing power. The particular reaction that Wächtershäuser singles out as most probable is the formation of pyrite, fool's gold, from ferrous sulfide and H_2S (or sulfide anion):

$$FeS + H_2S \rightarrow FeS_2 \text{ (pyrite)} + H_2$$
$$FeS + S^{2-} \rightarrow FeS_2 \text{ (pyrite)} + 2e^-$$

The reactants are available geologically, in submarine hot springs and elsewhere; and their combination is strongly favored, thanks largely to the extreme insolubility of pyrite. We have here a single reaction that is both highly reductive and strongly exergonic (energy-yielding), especially at high temperatures.

The free energy available from pyrite formation would be sufficient to reduce carbon dioxide or monoxide to formate, provided a coupling mechanism exists. This is important, for the key to a mineral origin of life is the reduction of CO_2 to organic metabolites in the absence of enzymes. Such chemistry would be futile were it to take place in solution, for its products would be lost by diffusion. Wächtershäuser envisages containment from the beginning, not by lipid membranes but by electrostatic association with a mineral surface. The earliest protometabolic reactions arose in a film on the surface of pyrite crystals, with their electronegative products held in place by the electropositive surface. This, incidentally, may be one reason why anionic substances (often with multiple charges) are so prominent in contemporary metabolism. Surface chemistry offers a second benefit, no less valuable than the first: condensation reactions, such as those required to produce macromolecules, are often thermodynamically favored when the products are bound to a surface, whereas in free solution hydrolysis prevails.

Metabolism must have begun with CO_2 fixation, which supplied the first stock of organic substances. Contemporary organisms employ several pathways to reduce CO_2, the most prominent of which is part of plant photosynthesis. This cannot have been the original route, for photosynthetic CO_2 fixation relies on the formation and transformation of sugars, which would not have existed on the prebiotic earth. Instead, Wächtershäuser makes a case for the gradual emergence of a reductive citric acid cycle, whose anionic products clung to the surface. "This establishes the first organized entity of life: a composite . . . sphere of metastable organic ligands around a growing cluster of pyrite. Their anionic bonding provides a trough of metastability in the overall cascade of redox energy flow—an intermediate within that energy flow, rather than the source and sink of the overall energy flow" (15). In other words, proto-organisms were not objects so much as dynamic chemical systems. The inputs are CO_2 and energy, the latter derived from pyrite formation; the output consists of organic molecules that remain associated with the surface for a time and eventually detach. Pyrite crystals supply both the reaction vessel and the primordial catalyst.

Much of the argument turns on the use of chemical and evolutionary principles to "retrodict" the metabolic pathways with which life began (where retrodiction, analogous to prediction, is the art of reconstructing the past from knowledge of the present). Starting with the iron-sulfur world of the original pyrite crystals, Wächtershäuser progressively evokes the biochemical core complete with thioesters, coenzymes, amino acids, sugars and redox carriers. Cellularization takes place with membranes generated by the developing system itself. The earliest protocells would have enclosed pyrite crystals and lived by pyrite formation, but later ones broke free of their inorganic heritage. Only with the advent of protocells can one imagine the invention of genes and proteins, and with it the transformation of chemistry into biology; but this crucial transition remains sketchy.

Wächtershäuser's ideas have drawn considerable attention, thanks both to their merit and to their feisty presentation. I myself find their general scope and direction attractive, because they mesh so well with my own bias that life began with energy flow and hauled itself up by its own bootstraps. But his hypothesis, like the others, soon finds itself in hot water. The proposal is so explicit that, if it were correct, one might by now expect verification of its central postulates; this has not happened. The formation of pyrite from FeS and H_2S does occur in nature, albeit slowly, and can drive some reductive reactions; but the crucial coupling to the fixation of CO_2 or CO has remained elusive, and with it the formation of an arsenal of acidic metabolites. Perhaps an additional mineral catalyst is required, or an altogether different source of energy and reducing power, such as the oxidation of ferrous iron mediated by ultraviolet light. I have reservations about the emergence of a network of organic reactions within that surface film, and find it difficult to envisage how scummy pyrite crystals can propagate newly-invented reactions, compete with one another and evolve as a population in the absence of some kind of heritable information. Many features, particularly the proposal for the emergence of chemiosmotic energy coupling, demand more credulity than this hardened skeptic can muster; even the thermodynamic premises now appear questionable (16). All the same, there may be a core of truth in the central idea. I would wager that Wächtershäuser, like Horatio Nelson before Copenhagen, sometimes puts his blind eye to the telescope, but that he points the blooming thing in the right direction.

A Bridge too Far?

In the beginning was the Word; so says the gospel of Saint John. Goethe's *Faust*, that prototypic modern man and scientist, thought otherwise: in the beginning was the Deed. Rephrased just a little, scholars still divide into those who seek the origin of life in information and those who look to energetics. Those who believe, as I do, that living organisms are autopoietic systems capable of evolution by variation and natural selection, must keep a foot in both camps and risk being scorned by both. But the definition really sharpens the issue: the question is not only how life arose on earth, but how nature generates organized material systems to which terms such as adaptation, function and purpose can be applied. Readers will have noted that this is still a free-wheeling inquiry, in which the few solid facts need not seriously impede the imagination; let me take advantage of what has, sadly, become a very rare privilege.

Granted that, as de Duve says, we are compelled by our calling to insist at all times on strictly naturalistic explanations; life must, therefore, have emerged from chemistry. Granted also that simple organic molecules were present at the beginning, in uncertain locations, diversity and abundance. Leave room for contingency, some rare chemical fluctuation that may have played a seminal role in the inception of living systems; and remember that you may be mistaken. With all that, I still cannot bring myself to believe that rudimentary organisms of any kind came about by the association of prefabricated organic molecules, born of purely chemical processes in their environment. Did life begin as a molecular collage? To my taste, that idea smacks of the reconstitution of life as we know it rather than its genesis *ab initio*. It overestimates what Harold Morowitz called the munificence of nature, her generosity in providing building blocks for free. It makes cellular organization an afterthought to molecular structure, and offers no foothold to autopoiesis. And it largely omits what I believe to be the ultimate wellspring of life, the thermodynamic drive of energy dissipation, creating mounting levels of structural order for natural selection to winnow. If it is true that life resides in organization rather than in substance, then what is left out of account is the heart of the mystery: the origin of biological order.

Scientists formulate hypotheses, not just at the conclusion of an inquiry but from its very outset. Karl Popper and Thomas Kuhn both taught that, absent a preconception of some sort, we do not know what questions to ask or even what facts to observe. The downside is that we will cling to an outworn hypothesis, well aware of its shortcomings, until a more credible alternative comes to hand. This, I suspect, is where the study of biopoiesis now stands: the past unburied, the future not yet born. I will also venture an opinion about where we should look. The hurdle is to understand, not the origin of organic molecules, but of systems that progressively come to display the characteristics of organisms: boundaries, metabolism, energy transduction, growth, heredity and evolution. This is hardly a startling or even original proposition, but its unapologetic holism makes it a minority view.

I hold, then, that cellular organization was not a codicil to the true origin of life, but part and parcel of it. That implies compartmentation of some kind (not necessarily lipid membranes) from the beginning. Biological order is dynamic, created and sustained by a continuous stream of energy, and that also must have been true all along. Therefore, a credible biopoietic theory will be one that generates mounting levels of complexity naturally, by providing the means to convert the flux of energy into organization. But energy dissipation can only carry life over the first jump; evolution is hamstrung until the emerging "functions" within the developing system have been codified in a "text" of some kind that can be transmitted, executed, altered, and put to the test of utility again and again. Nucleic acids or their precursors must have come on stage early, if not when the curtain rose. No satisfying scheme of this kind is presently on the books, and I have none to offer, I have only the strong hunch that there is much more to this mystery than is dreamt of in molecular philosophy.

It would be agreeable to conclude this book with a cheery fanfare about science closing in, slowly but surely, on the ultimate mystery; but the time for rosy rhetoric is not yet at hand. The origin of life appears to me as incomprehensible as ever, a matter for wonder but not for explication. Even the principles of biopoiesis still elude us, for reasons that are as much conceptual as technical. The physical sciences have been exceedingly successful in formulating universal laws on the basis of reproducible experiments, accurate measurements, and theories explicitly designed to be falsifiable. These commendable practices cannot be fully extrapolated to any historical subject, in which general laws constrain what is possible but do not determine the outcome. Here knowledge must be drawn from observation of what actually happened, and seldom can theory be directly confronted with reality. The origin of life is where these two ways of knowing collide. The approach from hard science starts with the supposition that physical laws exercise strong constraints on what was historically possible; therefore, even though one can never exclude the intervention of some unlikely but crucial happenstance, one should be able to arrive at a plausible account of how it could have happened. This, however, is not how matters have turned out. The range of permissible options is so broad, and the constraints so loose, that few scenarios can be firmly rejected; and when neither theory nor experiment set effective boundaries, hard science is stymied. The tools of "soft," historical science unfortunately offer no recourse: the trail is too cold, the traces too faint.

They tell a story of Max Delbrück, one of the pioneers of molecular genetics and the iconic inventor of DNA, whom I was privileged to meet during his later years at the California Institute of Technology. He had stopped reading papers on the origin of life, Max once observed; he would wait for someone to produce a recipe for the fabrication of life. So are we all waiting, not necessarily for a recipe but for new techniques of apprehending the utterly remote past. Without such a breakthrough, we can continue to reason, speculate and argue, but we cannot know. Unless we acquire novel and powerful methods of historical inquiry, science will effectively have reached a limit.

EPILOGUE

"All Faith is false, all Faith is true:
Truth is the shattered mirror strown
in myriad bits; while each believes
his little bit the whole to own."

Richard Francis Burton (1)

Schrödinger's riddle touches not only scientists and philosophers, but everyone who thinks about the world. What is life? How we answer that question must eventually impinge on the practice of medicine and law, influence what we teach our children, nudge the direction of economics and public policy, and color our attitude to man, God, and all ultimate concerns. To be sure, most of those who wonder about the meaning of life care chiefly about human life, and take little notice of the humble microbes that dwell in sea or soil and make up the foundation of all the biosphere. But these two levels of interest cannot be cleanly separated, for all known living things belong to a single tribe, related by composition, function and descent. The way of the cell is also the way of all flesh, ourselves included. By sticking closely to what science has learned about life in its simplest forms, I have carefully avoided the very issues that, to most people, matter most. But this was done for the sake of objectivity, not for lack of awareness or concern. In these final pages, let me gingerly prod the nettle and try to draw some personal meaning from scientific knowledge.

The bedrock premise of this book is that life is a material phenomenon, grounded in chemistry and physics. Life designates a quality, or property, of certain complex dynamic systems that persist by channeling through themselves streams of matter, energy and information. They have the unique capacity to reproduce themselves indefinitely, and arise on a millennial time-scale by the interplay of variation and selection that underlies biological evolution. Even the human mind emerges from the activities of the brain and represents a product of evolution, though these are matters of which we know little, and understand less. I know of no evidence for the existence of vital forces unique to living organisms, and their erratic history gives one no reason to believe that life's journey is directed toward a final destination in pursuit of a plan or purpose. If life is the creation of some cosmic mind or will, it has taken care to hide all material traces of its intervention. Now one can argue that so long as we confine our inquiries to the material side of life, material answers are all we can expect; they do not warrant the assumption that there are no other questions to be asked, with altogether different answers. Science alone may not be sufficient to make sense of the world, but I would insist that science is privileged; for of all the ways of questioning nature, science alone holds the promise of objective knowledge.

For the past century, biology has perched uneasily on the cusp of an ancient conflict: Evolution, the nature and origin of life, and the phenomenon of man mark a divide that separates the traditional and religious way of apprehending reality from the scientific one. One side beholds a world of purposeful order, manifesting the will of a divine architect who cherishes His creation and made man in His image. The other celebrates matter, energy, and universal laws tempered by the historical interplay of chance and necessity. The public has welcomed with glee the technological fruits of scientific endeavor, but recoils from its critical spirit and from its conclusions about the nature of things. If the polls are to be believed, the great majority of Americans keep the faith in a personal God, and more than half reject evolution; opinion in other industrialized nations is not quite so indifferent to the findings of science, but can hardly be said to embrace them. One has to wonder why, for all its achievements and manifest power, the spirit of science has traveled so badly. Scientists like to believe that better ways of engaging the public can bridge the gap, and the many ongoing projects to improve public understanding of science are all to the good. I suspect, however (as does Mary Midgley, whom I quoted in the preface on the subject of understanding), that the causes of disenchantment with science run much deeper. Science, and particularly the narrowly focused and reductionist science of the present day, is perceived as denying the world meaning; and without meaning humans cannot live.

I wonder, too, whether we touch here upon one of the causes of the spiritual malaise that so many sense in the modern world. We take pride in our superior understanding and our masterful technology, but it is plain that these were bought at substantial cost to human self-esteem. Not so long ago, Western man saw himself as God's own handiwork, dwelling upon the very pivot of creation. Contemporary humanity lives in much reduced circumstances, stuck on a small planet circling a mid-sized star, one among billions in an unremarkable galaxy, and there are billions more galaxies out there. The findings of biologists cut even closer to the bone. They compel us to admit that we humans, like all other organisms, are transient constellations of jostling molecules, brought forth by a mindless game of chance devoid of plan or intent. For anyone who takes science seriously, it becomes ever harder to believe that behind the appearances abides a cosmic mind that is even remotely comprehensible to us, or one that has the slightest concern for human welfare, personal or collective. In the absence of such a transcendent presence, many of the premises of civilization lose their historical moorings: that human life is sacred, that we can know right from wrong, that we are here to some purpose and that our little lives have larger meaning. We cannot go home again. But it is not at all self-evident that, absent a belief in powers greater than ourselves, a decent and civilized society can be sustained for long.

The universe revealed by science is under no obligation to be meaningful to mankind, and one can make a strong case that it is in fact utterly indifferent to us. That was Jacques Monod's view, compellingly argued thirty years ago in a slim but powerful missive entitled *Chance and Necessity* (2). A brilliant scientist who was also an accomplished cellist and a hero of the French Resistance to the Nazis, Monod took continental Europe by storm; curiously, the book seems to have made little impact on the more pragmatic English-speaking community. For Monod, the fundamental principle is that nature is "objective"; that is, nature has no motive, purpose or plan. He cuttingly dismissed as "animisms" the manifold ways in which men have, over untold millennia, sought to endow the world with direction, significance, mind or soul. Monod convinced himself that life arose entirely by chance, thanks to a constellation of infinitely improbable circumstances, and that its subsequent history displays nothing more purposeful than a succession of flukes. In a celebrated passage, Monod proclaimed that "the universe was not pregnant with life, nor the biosphere with man. Our number came up in the Monte Carlo game" (2). And he closed on a note of sombre yet romantic grandeur: "The ancient covenant is in pieces: man knows at last that he is alone in the universe's unfeeling immensity, out of which he emerged only by chance. His destiny is nowhere spelled out, nor is his duty. The kingdom above or the darkness below: it is for him to choose."

Monod's Gallic logic is hard to fault, but perhaps the underlying premises can be. It is by no means an indisputable fact that nature lacks direction, and that life in all its beauty and variety is wholly the product of accident and good luck. More than a few contemporary scientists believe that a tendency to self-organization is inherent in the physical universe, and that it underpins the emergence and progressive evolution of life. Incidentally, while science has but recently set foot upon this road, the notion of a creative and developing universe has long had standing among philosophers, such as Whitehead and Bergson. Monod would scathingly reject this attitude, insisting that emergent order is a delusion that we read into the universe in a desperate effort to deny the intolerable truth and to find in science reassurance or salvation. The search for salvation, in Mary Midgley's sense of a framework of ideas that dispels confusion and makes the world comprehensible, is surely a prominent theme of contemporary science; but that does not necessarily vitiate observations and arguments, particularly those that have arisen from physics. It appears to be the case that complex, dynamic systems generate structure and order spontaneously, in the absence of either intelligent design or Darwinian selection. This gives some degree of direction to the course of cosmic events, and allows one to envisage evolution as a triple game of variation, selection and natural law.

The universe suggested by this line of thought is not quite so bleak as Monod's frigid immensity; on the contrary, it is fruitful, exuberant, a place in which life finds a natural home. Yet it must be said that, as far as the human condition is concerned, the verdict of science remains substantially unchanged. As far as we know, humans (like every other living thing) represent a small and recent twig on that enormous tree of life, shaped in large measure by variation and selection. No intelligence, examining the bones and tools of *Homo habilis*, could have foreseen the Sistine Chapel, for it was not in any sense fore-ordained: evolution (cultural as well as physical) performs its wonders without intent, guidance or safety net. For better or for worse, mankind makes itself, and no one who wanders the globe can fail to be impressed by the sheer variety of choices that the human race has made. As our numbers and powers continue to mount, conflicts are certain to arise between all sorts of time-honored practices and beliefs (those of the West no less than those of Asia and Africa) and new necessities forced on us by science and technology. If a liveable world is to emerge from the race between sanity and catastrophe, we shall have to come to terms with the limits of our small planet; science must play a much larger role in shaping public policy than it does at present, particularly in matters of population and environment. But we must also find secular (or at least tolerant) soil in which to re-root those civilized values that sages have proclaimed time and again, usually in the name of one god or another.

For those of us who have outworn the ancient covenants between man and his gods, the search for meaning necessarily becomes personal rather than tribal. This, too, is hardly a strange road, for it has been trodden for centuries by Epicureans and Stoics, by Buddhists, Sufis, and all manner of free-thinkers; and many thoughtful moderns, including scientists, travel it today. I do not feel diminished by the discovery that we are all part of a vast biotic enterprise that brought forth consciousness, understanding, and morality from mindless chemistry. The great tree of life does not command my worship, but it surely evokes reverence and awe; and I would gladly surrender the illusion of dominion for the responsibilities of stewardship. For me, as for most humans past and present, the search for meaning remains unfinished, an aspiration rather than an achievement. And I am proud to walk in the company of Diogenes of Oenoanda who, nearly two thousand years ago, had this inscription engraved upon his tombstone (3): "Nothing to fear in God/Nothing to feel in death/Good can be attained/Evil can be endured."

I have come to think of science as a kind of game, whose object is to make rational sense of the world. Players are bound by strict rules: the imagination must ever be disciplined by reason, observation and experiment, and no cheating, please! It is the most engrossing game ever invented, one to which I and many others have happily dedicated our lives; and it has revealed much that is new, true and important. But we must never forget that the game of science is played on a board, and most of what matters most to human beings lies off the board. Science has little useful to say about good and evil, right and wrong, justice and oppression, and the strange ways of the human heart. Science can often explain what is happening, and it can sometimes forecast the future and distinguish wisdom from folly. But it provides no basis for ethical choice, nor the will to act. About what it means to be human, individual scientists often hold strong opinions; but science must be silent.

NOTES

Preface

1. Midgley, 1992, p 9.
2. Harold, 1986, p 563.

Acknowledgments

1. Tuchman, 1981, p 21.

Chapter 1: Schrödinger's Riddle

1. Schrödinger, 1944. The quotation will be found on p 68. Perutz' critical review (1987), appearing more than forty years later, testifies to the book's enduring influence.
2. Judson, 1979, reprinted and expanded 1996.
3. Weaver, 1948.
4. That life is not reducible to its physical constituents was forcefully argued by Polanyi, 1968.
5. Erwin Chargaff is a rarity among scientists, one who looks into the dark and reports what he sees there. See Chargaff, 1971, 1978, p 87.
6. Morowitz, 1968, reprinted 1979.

Chapter 2: The Quality of Life

1. Von Neumann is quoted in Pittendrigh, 1993. Order and organization will be employed interchangeably, but the two words emphasize different aspects of regularity. Order describes any methodical or predictable arrangement of elements. Wallpaper patterns are ordered, and so is a cell of *E. coli* whose composition is largely predictable from its identity. Organization is purposeful order; the wallpaper has order but not organization, the cell has both, and Rembrandt's Nightwatch has strong organization but little regularity. For an extensive discussion of the meanings of order, see Riedl, 1978.
2. See Pirie, 1938. Perret is quoted in Morowitz, 1987, p 186; I have not been able to locate the original source. For autopoietic systems see Varela et al., 1974; Fleischaker, 1988; and Margulis, 1993. Other sources are Dyson, 1988; Dulbecco, 1985, p 17; and Maynard Smith, 1986, p 7.
3. Crick, 1966, p 10.
4. See, among others, Simpson, 1963; Polanyi, 1968; Mayr, 1982; and Rosenberg, 1985.
5. The point is argued in detail by Mayr, 1982, 1988; and by Rosenberg, 1985.
6. Bonner, 1988, p IX; Hunter, 1996, Stent, 1968.

Chapter 3: Cells in Nature and in Theory

1. Beck, 1957, p 100.
2. Ford, 1981.
3. Schleiden, 1838; cited in Smith, 1976, p 213.
4. For more detailed accounts of the emergence of the cell concept see Baker, 1948, 1949; Smith, 1976; and Mayr, 1982.
5. Stanier and van Niel, 1962.
6. Margulis and Schwartz, 1998.
7. Protista, as Haeckel had it, or Protoctista (from the Greek for "first formed"). The latter usage, championed by Lynn Margulis in particular, has gained ground but is inappropriate for the present book, which is grounded in Carl Woese's conception of domains that supersede the traditional kingdoms. See below.
8. Woese and Fox, 1977; Fox et al., 1980; Woese et al., 1990.
9. Woese et al. 1990, designated the second of these domains Bacteria. To me and many others, Eubacteria seems preferable, so as to avoid confusion with the widespread usage of bacteria as a synonym for prokaryotes. For consistency with traditional practice, Eukarya will be spelled with a K.
10. Mayr, 1990. See also Margulis, 1992, 1996; Cavalier-Smith, 1993.
11. Wheelis et al., 1992.
12. Woese, 1987; Pace, 1997; Doolittle, 1998.
13. Gold, 1992; Whitman et al., 1998.
14. Stanier, 1970; de Duve, 1991; Alberts et al., 1994.
15. It is next to impossible to make any statement about biology that does not require qualification. *Nanochloron*, a eukaryotic member of the phytoplankton, has minute cells 2–3 micrometers in diameter. Nannobacteria, if they prove to be real, will be a tenth the size of conventional bacteria. On the other hand, the bacterium *Epulopiscium fishelsonii*, which inhabits the intestinal tract of the brown surgeonfish, is a giant by cellular standards: a single cell may reach 600 micrometers in length, with a diameter of 80 micrometers (Angert et al., 1993). Another such monster has been isolated from the sea off the coast of Namibia. Qualifications also apply to all the subsequent criteria, but have been omitted for brevity.
16. Jacob, 1982, p 11; 1973, p 121.
17. Woodger, 1967, p 291.

Chapter 4: Molecular Logic

1. Calvino, 1974, p 82. Quoted with permission of Harcourt, Inc. (HBT).
2. Lehninger et al., 1993, p 4; italics in the original.
3. Rensberger, 1997.
4. Thomas, 1974, p 171.
5. Harold, 1986.
6. Gold, 1992; Gould, 1996a.

7. Pace, 1991, 1997.

8. Stent, 1968; Chargaff, 1971; Judson, 1979.

9. Crick, 1970.

10. Jacob, 1973, p 251.

11. Bray, 1995.

12. Hyman and Karsenti, 1996; Newport, 1987.

13. Lederberg, 1966.

14. Monod, 1971, p 96.

15. Gould, 1989; Ohta and Kreitman, 1996

Chapter 5: A (almost) Comprehensible Cell

1. Eiseley, 1946, pp 195 and 202.

2. Neidhardt et al., 1996. For less intimidating surveys see Neidhardt et al., 1990, and White, 1995.

3. Perutz, 1986; see also Hunter, 1996.

4. Jacob, 1973, p 313.

5. Dawkins, 1982, 1995, 1996.

6. See also Goodwin, 1986, 1994; Harold, 1990, 1995; Nijhout, 1990.

7. Blattner et al., 1997.

8. Mushegian and Koonin, 1996; Maniloff, 1996. A very similar size has recently been inferred from experiments.

9. Stent, 1978.

10. Riley, 1993; Riley and Labedan, 1996.

11. Bray, 1995.

12. Neidhardt and Savageau, 1996.

13. Morowitz, 1968.

14. Landman, 1991.

15. Ho, 1993, 1994.

16. Mittenthal et al., 1993.

17. Macnab, 1996.

18. Kellenberger, 1990.

19. Kuhn, 1970. It has been argued that Kuhnian revolutions, whose hallmark is the replacement of an established paradigm by a new and incompatible one, do not occur in biology (Wilkins, 1996). The chemiosmotic revolution is one unequivocal example, and we were fully aware of it at the time (Harold, 1978).

20. Essentially, though not in every detail, as laid out in Mitchell, 1966.

21. Mitchell, 1979; Harold, 1986; Nicholls and Ferguson, 1992.

22. Harold, 1977, 1986; Harold and Maloney, 1996.

23. Adler, 1966.

24. Koshland, 1977.

25. Parkinson, 1993; Blair, 1995; Stock and Surette, 1996.

26. Bray and Bourrett, 1995; Spiro et al., 1997; Bray et al., 1998. Alon et al., 1999.

27. Bray, 1995, 1998; Maddock et al., 1993.
28. Ritter, 1919.
29. Matthews, 1993.
30. Norris et al., 1996; Nanninga, 1998; Shapiro and Losick, 1997.
31. Laszlo, 1972; Riedl, 1978.
32. Ochman and Lawrence, 1996.

Chapter 6: It Takes A Cell To Make A Cell

1. Gilbert, 1991.
2. Katz, 1986.
3. Arnone and Davidson, 1997.
4. See Goodwin, 1986, 1994; Harold, 1990, 1995.
5. Waddington, 1957. No two scholars employ epigenesis in exactly the same sense. The present usage is a broad one, covering all the processes by which genetic information is translated into the structure and functions of a cell. Spatial markers, modifications that alter gene expression and even some metabolic states can be transmitted from one generation to the next independently of the genes, and constitute a distinct channel of epigenetic inheritance. The cellular and physiological context within which genetic information is expressed, and which guides the course of development, makes up the epigenetic landscape.
6. Reviewed by Donachie, 1993; Rothfield, 1994; Koch, 1995; Cooper, 1996; Lutkenhaus and Mukherjee, 1996; Lutkenhaus and Addinall, 1997; Nanninga, 1998.
7. Park, 1996.
8. Jacob et al. 1963.
9. See Koch et al., 1981; Koch, 1985, 1988, 1992, 1995.
10. Cooper, 1996.
11. Is it really true that prokaryotes lack mechanoproteins and a cytoskeleton? Generally speaking yes, but one can argue that the Z-ring represents a transient mitotic apparatus, and there is some evidence that a motor protein is involved in chromosome segregation.
12. See Park, 1996; Höltje, 1998; Nanninga, 1998.
13. Höltje, 1996, 1998. In principle, such a multifunctional complex of peptidoglycan hydrolase and synthase could replicate the entire sacculus, allowing bacterial shape to be passed from one generation to the next by direct inheritance of structural information without the intervention of soap-bubble physics. In my opinion, this is unlikely; the fact that spheroplasts can regenerate the original cell shape suggests, at least, that any such template mechanism cannot be obligatory.
14. Donachie, 1993; Donachie et al., 1995.
15. Rothfield and Zhao, 1996; Wheeler and Shapiro, 1997.
16. For current ideas on how cells find their midpoint see Koch and Höltje, 1995; Nanninga, 1998; Raskin and de Boer, 1999.

17. Lutkenhaus and Mukherjee, 1996; Erickson, 1997; Lutkenhaus and Addinall, 1997.

18. See Dennett, 1995, p 114 for a particularly lucid discussion of what is meant when DNA is described as computer software.

19. Shapiro and Losick, 1997.

20. See Rothfield, 1994; Norris et al., 1996; Shapiro and Losick, 1997.

Chapter 7: Morphogenesis: Where Form and Function Meet

1. Russell, 1916, foreword.

2. Goodwin, 1993.

3. Harold, 1990, 1994, 1995, 1997, 1999.

4. Ingber, 1993, 1998.

5. Thompson, 1961; the quote will be found on p 11. For a more recent survey of the physical basis of natural patterns see Ball, 1999.

6. Reviewed in Harold, 1990; Heath, 1990; Wessels and Meinhardt, 1994; Gow and Gadd, 1995.

7. Harold et al., 1996; for a thoughtful review of the implications see Money, 1997.

8. Harold, 1997. The calcium hypothesis has a history of more than thirty years, beginning with the pioneering researches of Lionel Jaffe, Richard Nuccitelli and Kenneth Robinson (reviewed by Jaffe, 1981). For its application to apical growth see Harold, 1990, 1997; Hyde and Heath, 1997; and Taylor and Hepler, 1997.

9. Bartnicki-Garcia et al., 1989, 1995; Riquelme et al., 1998.

10. Mitchell, 1962.

11. See Kropf, 1992; Fowler and Quatrano, 1997; Alessa and Kropf, 1999.

12. Goodwin, 1994, p 44.

13. Roemer et al., 1996; Mata and Nurse, 1998.

14. For recent surveys see Bray, 1992; Bray and White, 1988; Condeelis, 1993; Lee et al, 1993; Grebecki, 1994; Lauffenburger and Horwitz, 1996.

15. Theriot, 1996.

16. Fulton, 1981.

17. See Frankel 1989, 1990, 1992, 1997 (the quote comes from Frankel, 1989, p 8); Grimes, 1990, Grimes and Aufderheide, 1991.

18. Sonneborn, 1970.

19. Frankel and Whiteley, 1993.

20. For accessible introductions see Prigogine and Stengers, 1984; Lewin, 1992; Ball, 1999.

21. Goodwin 1986, 1993, 1994, 1997; Harrison 1993.

22. Turing, 1952.

23. Meinhardt, 1982; Nüsslein-Volhardt, 1996; Lawrence and Struhl, 1996; Neumann and Cohen, 1997.

24. Brandts and Trainor, 1990 a, b; Brandts and Totafurno, 1997.

25. See Goodwin and Trainor, 1985; Brie're and Goodwin, 1988; Good-

win, 1994, 1997 (The quote is from Goodwin, 1994, p 111). For an alternative interpretation see Harrison, 1993; Dumais and Harrison, 2000.

26. Harold, 1997.

27. See Goodwin, 1994; Webster and Goodwin, 1996; Gilbert et al., 1996; Harrison, 1993.

28. Sapp, 1998.

Chapter 8: The Advance of the Microbes

1. Stewart and Golubitsky, 1992, p 127.

2. Woese, 1998b.

3. Gould, 1994.

4. Fermor, 1958.

5. Reviewed by Woese, 1987; Pace, 1997; Brown and Doolittle, 1997.

6. Woese et al., 1990; Wheelis et al., 1992. The philosophical underpinnings of the molecular approach to phylogeny have been vigorously defended by Woese, 1998b.

7. For the characteristics of Archaea and Eubacteria see Doolittle, 1996; Brown and Doolittle, 1997; Olsen and Woese, 1997, and Pace, 1997.

8. On cell walls see Kandler and König, 1993, on membrane lipids Langworthy and Pond, 1986.

9. Iwabe et al., 1989; Gogarten et al., 1989. See also Gogarten et al., 1996, for a clear discussion of the trouble with roots. It is fitting to recall here the pioneering work of Margaret Dayhoff and her colleagues who, more than twenty years ago, recognized the phylogenetic uses of gene duplication.

10. Phylogenetic trees based on protein sequences are carefully considered by Doolittle, 1996; Brown and Doolittle, 1997, and (from a radically different perspective) by Gupta, 1998.

11. Pace, 1997, argues the case for a geochemical origin of energy metabolism. See also Castresana and Moreira, 1999, for the phylogeny of redox chains, Gogarten et al., 1996 for ATPases.

12. Woese, 1998a, reconsiders the nature of the universal ancestor.

13. One should keep in mind that many lithotrophic organisms utilize oxygen, and are thus indirectly sustained by sunlight. All eukaryotes require oxygen, if not as an ultimate electron acceptor then as the oxidant in certain biosyntheses. Gould, 1996a.

14. Hyperthermophilic ancestors? See Pace, 1991, 1997; Stetter, 1994 in favor; Forterre, 1996 and Forterre and Philippe, 1999, opposed to the idea.

15. The use of protein clocks to probe deep time is described by Doolittle et al., 1996, quickly revised by Feng et al., 1997.

16. Fossil bacteria are authoritatively described by Schopf, 1993, 1996; Hayes, 1996, summarizes the geochemical data from Greenland.

17. See Koch, 1994.

18. The flagellar apparatus of Archaea is discussed by Jarrel et al., 1996,

and Faguy and Jarrell, 1999. Kakinuma, 1998 considers the ATPases of enterococci.

19. There is a growing literature of dissent from students of protein phylogeny. See Forterre and Philippe, 1999; Brown and Doolittle, 1997; Doolittle, 1999 and the forceful argument by Gupta, 1998.

20. Three domains or two empires? For the latest salvos in this continuing engagement (begun in Chapter 3) see Mayr, 1998; the rejoinder by Woese, 1998b, and the most recent universal taxonomy by Cavalier Smith, 1998. I continue to adhere to the paradigm of three domains, but with growing reservations. A repositioning of the root of the universal tree, putting Eubacteria and Archaea together on one side and Eukarya on the other, is all that divides the two viewpoints (albeit not the protagonists).

21. Margulis, 1970, 1993, 1996; Dyer and Obar, 1994; Stanier, 1970.

22. Reviewed by Gray and Doolittle, 1982; Woese, 1987; and Gray et al., 1999. See also the pioneering article by John and Whatley, 1975.

23. Mitochondria and chloroplasts are probably monophyletic: see Cavalier Smith, 1992; Gray and Spencer, 1996; and Gray et al., 1999.

24. Cavalier Smith, 1987, 1993; Cavalier-Smith and Chao, 1996. The status of the Archezoa is reconsidered in Cavalier-Smith, 1998; Keeling, 1998; and Doolittle, 1999.

25. The molecular perspective is represented by Woese, 1987, Olsen and Woese, 1997; Pace, 1997; Brown and Doolittle, 1997; and Doolittle, 1996, 1999.

26. See Erickson, 1997; Burns, 1998, and the cover of that issue of Nature.

27. Prokaryotic mergers are as attractive just now as corporate ones are! See Sogin et al., 1996; Brown and Doolittle, 1997; and Gupta, 1998. Syntrophic association is promoted by Martin and Müller, 1998; and Moreira and Lopez Garcia, 1998.

28. Margulis and Schwartz, 1998; Margulis et al., 1990.

29. For recent discussions of protistan phylogeny see Sogin, 1994; Sogin et al., 1996; Cavalier-Smith, 1993, 1998. Also Margulis, 1996.

30. For *Giardia* and its relatives see Gillin et al., 1996; and Cavalier-Smith and Chao, 1996.

31. The nature and status of the Archezoa is presently a matter of growing debate. See Keeling, 1998; Cavalier Smith, 1998; and Doolittle, 1999.

32. See Knoll, 1992, 1999; Sogin, 1994; Runnegar, 1994. Also Whatley, 1993, concerning the membranes that surround chloroplasts. *Grypania* is described by Han and Runnegar, 1992.

33. This subject will be re-visited in Chapter 9. See Cavalier Smith, 1987, 1988; de Duve, 1991; Maynard Smith and Szathmàry, 1995.

Chapter 9: By Descent With Modification

1. Medawar, 1982, p 46.
2. This theme has been forcefully expounded by Dennett, 1995.

3. For a summary and history of the modern synthesis see Mayr, 1982. For a systematic and accessible critique see Gould, 1982, and Gould and Eldredge, 1993; the quotations come from these articles.

4. Gilbert et al., 1996.

5. "Natural selection" is used here in a broad sense, to include what Darwin called "sexual selection" and other modes of selection for reproductive success.

6. Dawkins, 1982, 1995, 1996; Dennett, 1995.

7. See Mayr, 1997, for a magisterial disquisition on the objects upon which natural selection acts.

8. This archaic but useful term has recently been revived by E. O. Wilson. The Oxford English Dictionary defines it as "the accordance of two or more inductions drawn from different groups of phenomena."

9. Gould, 1997. Articles by Gould in the June 12 and June 26, 1997 issues of the *New York Review of Books*, and reply by Dennett.

10. In addition to the forceful argument of Gilbert et al., 1996, see Raff, 1996; and the much earlier work of Riedl, 1978, who emphasized the importance of developmental constraints in evolution.

11. Major contributions come from Goodwin, 1994; Webster and Goodwin, 1981, 1996; and Kauffman, 1993, 1995.

12. The theme of evolutionary transitions has been fully developed by Szathmàry and Maynard Smith, 1995, and by Maynard Smith and Szathmàry, 1995. See also the earlier treatment by Bonner, 1988.

13. Cavalier-Smith's views have themselves evolved somewhat over time; see Cavalier-Smith, 1987, 1988, 1991, 1993; Cavalier-Smith and Chao, 1996; and also Keeling, 1998.

14. Shapiro, 1997. A decade ago it was reported that many bacterial mutations are "directed," in the sense that they arise in response to some particular stress and relieve that stress. This, if true, would squarely contravene the Darwinian view of evolution. However, it now appears that these apparently adaptive mutations arise, in fact, as the result of random and orthodox mechanisms. The debate continues.

15. There is a huge literature on molecular evolution, much less about the evolution of cellular systems. See Kirschner, 1992; Kirschner and Gerhart, 1998; Maynard Smith and Szathmàry, 1995; several contributions to a recent symposium (e.g. Mitchison, 1995; Nasmyth, 1995); and Becker and Melkonian, 1996.

16. Behe, 1996; see also the rebuttal by Coyne, 1996.

17. See recent reviews by Kidwell, 1993; Syvanen, 1994, and by Doolittle, 1999. Also papers by Katz, 1996, and by Lawrence and Ochman, 1998.

18. See Margulis and Fester, 1991; Margulis, 1993; and also Sapp, 1994, for a historical overview.

19. Cyanelles and *Mixotricha* are described in Margulis, 1993. The purple protist was discovered by Fenchel and Bernard, 1993, the hydrogenosome with

a genome by Akhmanova et al., 1998. An article by Jeon in Margulis and Fester, 1991, describes his experiences with bacterial endosymbionts in amoebas.

20. Besides the comprehensive survey by Jablonka and Lamb, 1995, microbiologists will find much of interest in the articles by Landman, 1991, and by Preer, 1993.

21. Carlile, 1982 considers bacteria to be r-strategists, protists are K-strategists. These arcane terms come from a mathematical formulation often employed in ecology. Where the Archaea fit into his scheme is not clear.

22. See Kirschner and Gerhart, 1998, and also the accompanying comment by West Eberhard. The influence of Michael Conrad's pioneering article of 1990 is evident and fully acknowledged.

23. Simon, 1962 illustrated the virtues of modularity with the parable of two watchmakers; they were equally skilled, yet Hora prospered while Tempus went bankrupt. Tempus, it turns out, made every watch as a unit; whenever he put his work down to serve a customer the unfinished watch fell apart and all his labor was lost. Hora made his of subassemblies or modules that were stable enough to survive interruption. Modularity is also a major theme in Riedl's book, 1978.

24. See Gould and Eldredge, 1993; Knoll, 1992; and Benton, 1995.

25. Gould, 1996b. The quotes will be found on pages 28, 29.

26. Leopold, 1949, p 224.

Chapter 10: So What Is Life?

1. Chargaff, 1971.

2. Harold Morowitz recounts several in Cosmic Joy and Local Pain, 1987. In one version, the seeker comes to the great guru in his mountain fastness and pleads "What is Life?" The sage answers, "Life is a fountain." The supplicant, not surprisingly, is annoyed: "I have traveled halfway around the world, spent a fortune, risked my life, and all you can tell me is that life is a fountain?" "All right, my son," says the guru, "for you, life is not a fountain."

3. Bronowski, 1978, p 134.

4. Dinesen, 1934, The dreamers.

5. The relevance of Kant's philosophy to issues in contemporary biology is lucidly discussed by Webster and Goodwin, 1981, 1996, and Wicken, 1987.

6. Lewin, 1992, offers a readable introduction to complexity, Nicolis and Prigogine, 1989, a more technical one. The popular treatment by Prigogine and Stengers, 1984, is also helpful.

7. Stein, W., quoted by E. F. Yates in Boyd and Noble, 1993.

8. Rosen, 1985, 1991, makes very few concessions to the general reader, but has important points to make. The quote comes from Rosen, 1991, p 244.

9. For a new approach to the perennial mystery of protein folding see Dill and Chan, 1997.

10. Kauffman, 1993, 1995.

11. Conrad, 1990, p 79.

12. I have drawn particularly on the following books and articles: Wicken, 1987, Weber et al., 1988; Goodwin et al., 1989; Conrad, 1990; Saunders, 1993; Fontana and Buss, 1994; Weber and Depew, 1996; and Depew and Weber, 1998.

13. For an accessible introduction to the second law see Atkins, 1984. The fundamental work on the relationship of that law to biology is Morowitz, 1968. The most useful general treatment of bioenergetics is, of course, that by Harold, 1986.

14. See Wicken, 1987; Weber et al., 1988; Weber and Depew, 1996; Berry, 1995.

15. Wicken, 1987, pp 72 and 136.

16. Lenton, 1998.

17. See, for example, Wicken, 1987; Maynard Smith and Szathmàry, 1995.

18. For instance, Rosenberg, 1985, thinks otherwise.

19. For a discussion of this and other self-reproducing entities see Poundstone, 1985.

20. Sagan et al., 1993.

21. See, for example, Gould, 1996b, and his earlier book Wonderful Life.

Chapter 11: Searching for the Beginning

1. In recent years there has been a spate of books that deal wholly or largely with the origin of life. Among these see Shapiro, 1986; Küppers, 1990; de Duve, 1991; Morowitz, 1992; Eigen, 1992; de Duve, 1995; Maynard Smith and Szathmàry, 1995; Zubay, 1996; and Davies, 1999. Articles accessible to the general reader are due to Horgan, 1991; Orgel, 1994; and Bernstein et al., 1999.

2. Crick, 1981.

3. Holzman, 1999.

4. See Pace, 1991; Lazcano and Miller, 1996.

5. The odds are carefully weighed by Shapiro, 1986, whose tone of respectful but implacable skepticism matches my own mood.

6. For recent discussions of the primordial soup see Orgel, 1994; Maynard Smith and Szathmàry, 1995; Lazcano and Miller 1996; and Zubay, 1996.

7. de Duve, 1991, 1995; Kauffman, 1993, 1995.

8. Among those now sounding notes of caution are Orgel, 1998; Lazcano and Miller, 1996; and Keefe and Miller, 1995. Shapiro, 1999, remains deeply skeptical.

9. A partial list must include Gilbert, 1986; Joyce, 1989; Eigen, 1992; Elitzur, 1994; Maynard Smith and Szathmàry, 1995; Lifson, 1997; Gilbert and de Souza, 1999, and other contributors to The RNA World (Gesteland, Cech and Atkins, 1999).

10. Reviewed by Orgel, 1998; and Eschenmoser, 1999.

11. The case for membranes and energetics is forcefully made by Morowitz,

1992; and by Deamer, 1997. Even those who insist that life began with informational molecules recognize that enclosure must have come early (e.g., Eigen, 1992; Lifson, 1997; Orgel, 1998).

12. Concerning the evolution of energy transduction see Koch and Schmidt, 1991; and Deamer, 1997. The idea of inside out cells (obcells) was first mooted by Günter Blobel in 1980, and plays a large role in the thinking of Cavalier-Smith, 1987, and of Maynard Smith and Szathmàry, 1995.

13. For a sense of how this concept originated and what it now means see Gilbert, 1986; Joyce, 1989; Orgel, 1998; Jeffares et al., 1998; Bartel and Unrau, 1999; and the second edition of The RNA World (Gesteland, Cech and Atkins, 1999).

14. See Pace, 1991, and for a contrary opinion Miller and Lazcano, 1995.

15. Wächtershäuser, 1988, 1992, 1994, 1997. An introductory review has been prepared by Maden, 1995. The quotations come from the 1994 and 1997 articles.

16. See Schoonen et al., 1999.

Epilogue

1. Burton, 1880, p 42.
2. Monod, 1971, Quotations from p 145–146 and 180.
3. Muller, 1958, p 157.

REFERENCES

Adler, J. (1966) Chemotaxis in bacteria. Science *153*:708–716.

Akhmanova, A., Vonken, F., van Alen, T., van Hoek, A., Boxma, B., Vogels, G., Veenhuist, M., and Hackstein, J. H. P. (1998) A hydrogenosome with a genome. Nature *396*:527–528.

Alberts, B., Bray, D., Lewis, J., Raff, M., Roberts, K., and Watson, J. D. (1994) Molecular Biology of the Cell, 3rd ed. Garland Publishing, New York.

Alessa, L. and Kropf, D. L. (1999) F-actin marks the rhizoid pole in living *Pelvetia compressa* zygotes. Development *126*:201–209.

Alon, V., Surette, M. G., Barkal, N., and Leibler, S. (1999) Robustness in bacterial chemotaxis. Nature *397*:168–171.

Angert, E. R., Clements, K. D., and Pace, N. R. (1993) The largest bacterium. Nature *362*:239–241.

Arnone, M. I. and Davidson, E. H. (1997) The hardwiring of development: organization and function of genomic regulatory systems. Development *124*:1851–1864.

Atkins, P. W. (1984) The Second Law. Scientific American Library, W. H. Freeman and Company, New York.

Baker, J. R. (1948) The cell theory: A restatement, history and critique. Part I. Quarterly Journal of Microscopic Science *89*:103–125.

Baker, J. R. (1949) The cell theory: A restatement, history and critique. Part II. Quarterly Journal of Microscopic Science *90*:87–108.

Ball, P. (1999) The Self-Made Tapestry. Oxford University Press, Oxford and New York.

Bartel, D. P. and Unrau, P. J. (1999) Constructing an RNA world. Trends in Biochemical Sciences *24*:M9–M13.

Bartnicki-Garcia, S., Bartnicki, D. D., and Gierz, G. (1995) Determinants of fungal cell wall morphology: the vesicle supply center. Canadian Journal of Botany *73*, Supplement 1, S372–S378.

Bartnicki-Garcia, S., Hergert, F., and Gierz, G. (1989) Computer simulation of fungal morphogenesis and the mathematical basis of hyphal tip growth. Protoplasma *153*:46–57.

Beck, W. S. (1957) Modern Science and the Nature of Life. Harcourt, Brace & Co., New York.

Becker, B., and Melkonian, M. (1996) The secretory pathway of protists: spatial and functional organization and evolution. Microbiological Reviews 60:697–721.

Behe, M. J. (1996) Darwin's Black Box: The Biochemical Challenge to Evolution. Free Press/Simon and Schuster, New York.

Benton, M. J. (1995) Diversification and extinction in the history of life. Science 268:52–58.

Bernstein, M. D., Sandford, S. A. and Allamandola, L. J. (1999) Life's far-flung raw materials. Scientific American, July, 42–49.

Berry, S. (1995) Entropy, irreversibility and evolution. Journal of Theoretical Biology 175:197–202.

Blair, D. F. (1995) How bacteria sense and swim. Annual Review of Microbiology 49:489–522.

Blattner, F. R., Plunkett, G. III, Bloch, C. A., Perna, N. T., Burland, V., Riley, M., Collado-Vides, J. Glasner, J. D., Rode, C. K., Mayhew, G. F., Gregor, J., Davis, N. W., Kirkpatrick, H. A., Goeden, M. A., Rose, D. J., Mau, B., and Shao, Y. (1997) The complete genome sequence of *Escherichia coli* K12. Science 277:1453–1462.

Bonner, J. T. (1988) The Evolution of Complexity by Natural Selection. Princeton University Press, Princeton.

Boyd, C. A. R. and Noble, D. (eds.) (1993) The Logic of Life: The Challenge of Integrative Physiology. Oxford University Press, New York.

Brandts, W. A. M. and Totafurno, J. (1997) Vector field models of morphogenesis. In: Physical Theory in Biology: Foundations and Explorations (C. J. Lumsden, W. A. Brandts and L. E. H. Trainor, eds.) 107–140. World Scientific, Singapore, New Jersey, London and Hong Kong.

Brandts, W. A. M. and Trainor, L. E. H. (1990a) A non-linear field model of pattern formation: intercalation in morphallactic regulation. Journal of Theoretical Biology 146:37–56.

Brandts, W. A. M. and Trainor, L. E. H. (1990b) A non-linear field model of pattern formation: application to intracellular pattern formation in *Tetrahymena*. Journal of Theoretical Biology 146:57–86.

Bray, D. (1992) Cell Movements. Garland Publishing, Inc., New York and London.

Bray, D. (1995) Protein molecules as computational elements in living cells. Nature 376:307–312.

Bray, D. (1998) Signaling complexes: Biophysical constraints on intracellular communication. Annual Review of Biophysics and Biophysical Chemistry 27:59–75.

Bray, D. and Bourrett, R. B. (1995) Computer analysis of the binding reactions leading to a transmembrane receptor-linked multiprotein complex involved in bacterial chemotaxis. Molecular Biology of the Cell 6:1367–1380.

Bray, D. and White J. G. (1988) Cortical flow in animal cells. Science *239*: 883–888.

Bray, D., Levin, M. D., and Morton-Firth, C. J. (1998) Receptor clustering as a cellular mechanism to control sensitivity. Nature *393*:85–88.

Brie're, C. and Goodwin, B. C. (1988) Geometry and dynamics of tip morphogenesis in *Acetabularia*. Journal of Theoretical Biology *131*:461–475.

Bronowski, J. (1978) The Common Sense of Science. Harvard University Press, Cambridge, MA.

Brown, J. R. and Doolittle, W. F. (1997) Archaea and the prokaryote-eukaryote transition. Microbiological Reviews *61*:456–502.

Burns, R. (1998) Synchronized division proteins. Nature *391*:121–122.

Burton, R. F. (1880) The Kasidah of Haji Abdu El-Yezdi. Willey Books, New York, 1944.

Calvino, I. (1974) Invisible Cities. (Translated by W. Weaver). Harcourt Brace Jovanovich, New York.

Carlile, M. (1982) Prokaryotes and eukaryotes: strategies and successes. Trends in Biochemical Sciences *7*:128–130.

Castresana, J. and Moreira, D. (1999) Respiratory chains in the last common ancestor of living organisms. Journal of Molecular Evolution *49*: 453–460.

Cavalier-Smith, T. (1987) The origin of cells: a symbiosis between genes, catalysts and membranes. Cold Spring Harbor Symposia on Quantitative Biology *52*:805–824.

Cavalier-Smith, T. (1988) Origin of the cell nucleus. BioEssays *9*:72–78.

Cavalier-Smith, T. (1991) Cell diversification in heterotrophic flagellates. In: The Biology of Free-Living Heterotrophic Flagellates (D. J. Patterson and J. Larsen, eds.). pp 113–131. Clarendon Press, Oxford.

Cavalier-Smith, T. (1992) The number of symbiotic origins of organelles. BioSystems *28*:91–106.

Cavalier-Smith, T. (1993) Kingdom protozoa and its 18 phyla. Microbiological Reviews *57*:953-994.

Cavalier-Smith, T. (1998) A revised six-kingdom system of life. Biological Reviews *73*:203–266.

Cavalier-Smith, T. and Chao, E. E. (1996) Molecular phylogeny of the free-living archezoan *Trepomonas agilis* and the nature of the first eukaryote. Journal of Molecular Evolution *43*:551–562.

Chargaff, E. (1971) Preface to a grammar of biology. Science *172*:637–642.

Chargaff, E. (1978). Heraclitean Fire: Sketches from a Life before Nature. Rockefeller University Press, New York.

Condeelis, J. (1993) Life at the leading edge: the formation of cell protrusions. Annual Review of Cell Biology *9*:411–444.

Conrad, M. (1990) The geometry of evolution. BioSystems *24*:61–81.

Cooper, S. (1996) Segregation of cell structures. In: Neidhardt et al., op. cit, pp 1652–1661.

Coyne, J. A. (1996) God in the details (a review of Darwin's Black Box). Nature *383*:227–228.

Crick, F. H. C. (1966) Of Molecules and Men. University of Washington Press, Seattle.

Crick, F. H. C. (1970) Central dogma of molecular biology. Nature *227*:561–563.

Crick, F. H. C. (1981) Life Itself: Its Origin and Nature. Simon and Schuster, New York.

Davies, P. (1999) The Fifth Miracle. Simon and Schuster, New York.

Dawkins, R. (1982) The Extended Phenotype. Oxford University Press, Oxford, New York.

Dawkins, R. (1995) River Out of Eden. Harper Collins Publishers, New York.

Dawkins, R. (1996) Climbing Mount Improbable. W. W. Norton and Co., New York, London.

de Duve, C. (1984) A Guided Tour of the Living Cell. Scientific American Library, W. H. Freeman and Co., New York.

de Duve, C. (1991) Blueprint for a Cell: The Nature and Origin of Life. Neil Patterson Publishers, Burlington, North Carolina.

de Duve, C. (1995) Vital Dust: Life as a Cosmic Imperative. Basic Books, New York.

Deamer, D. W. (1997) The first living systems: a bioenergetic perspective. Microbiology and Molecular Biology Reviews *61*:239–261.

Dennett, D. C. (1995) Darwin's Dangerous Idea: Evolution and the Meanings of Life. Simon and Schuster, New York.

Depew, D. J. and Weber, B. H. (1998) What does natural selection have to be like to work with self organization? Cybernetics and Human Knowing *5*:18–31.

Dill, K. A. and Chan, H. S. (1997) From Levinthal pathways to funnels. Nature Structural Biology *4*:10–19.

Dinesen, I. (1934) Seven Gothic Tales.

Donachie, W. D. (1993) The cell cycle of *Escherichia coli*. Annual Review of Microbiology *47*:199–230.

Donachie, W. D., Addinall, S., and Begg, K. (1995) Cell shape and chromosome partition in prokaryotes or, Why *E. coli* is rod shaped and haploid. BioEssays *17*:569–576.

Doolittle R. F., Feng, D. F., Tsang, S., Cho, G., and Little, E. (1996) Determining divergence times with a protein clock. Science *221*:470–477.

Doolittle, R. F. (1998) Microbial genomes opened up. Nature *392*:339–342.

Doolittle, W. F. (1996) Some aspects of the biology of cells and their possible evolutionary significance. In: Evolution of Microbial Life (D. Mc L. Roberts, P. Sharp, G. Alderson, and M. Collins, eds.). Society for General Microbiology Symposium *54*:1–22.

Doolittle, W. F. (1999) Phylogenetic classification and the universal tree. Science *284*:2124–2128.

Dulbecco, R. (1985) The Design of Life, Yale University Press, New Haven and London.

Dumais, J. and Harrison, L. G. (2000) Whorl morphogenesis in the dasycladalean algae: the pattern formation viewpoint. Philosophical Transactions of the Royal Society of London, Series B 355:281–306.

Dyer, B. D. and Obar, R. A. (1994) Tracing the History of the Eukaryotic Cell. Columbia University Press, New York.

Dyson, F. (1988) Infinite in All Directions. Harper and Row, New York.

Eigen, M. (1992) Steps toward Life: A Perspective on Evolution. Oxford University Press, New York.

Eiseley, L. (1946) The Immense Journey. Random House, New York.

Elitzur, A. C. (1994) Let there be life: thermodynamic reflections on biogenesis and evolution. Journal of Theoretical Biology 168:429–459.

Erickson, H. P. (1997) FtsZ, a tubulin homologue in prokaryotic cell division. Trends in Cell Biology 7:362–367.

Eschenmoser, A. (1999) Chemical etiology of nucleic acid structure. Science 284:2118–2124.

Faguy, D. M. and Jarrell, K. F. (1999) A twisted tale: The origin and evolution of motility and chemotaxis in prokaryotes. Microbiology 145:279–281.

Fenchel, T. and Bernard, C. (1993) A purple protist. Nature 362:300.

Feng, D. D., Cho, G. and Doolittle, R. F. (1997) Determining divergence times with a protein clock: update and reevaluation. Proceedings of the National Academy of Sciences USA 94:13028–13033.

Fermor, P. (1958). Mani: Travels in the Southern Peloponnese. London, Murray.

Fleischaker, G. R. (1988) Autopoiesis: the status of its systems logic. BioSystems 22:37–49.

Fontana, W. and Buss, L. W. (1994) What would be conserved if "the tape were played twice?" Proceedings of the National Academy of Sciences U.S.A. 91:757–761.

Ford, B. J. (1981) Leeuwenhoek's specimens discovered after 307 years. Nature 292:407.

Forterre, P. (1996) A hot topic: the origin of hyperthermophiles. Cell 85:789–792.

Forterre, P. and Philippe, H. (1999) Where is the root of the universal tree of life? BioEssays 21:871–879.

Fowler, J. E. and Quatrano, R. S. (1997) Plant cell morphogenesis: plasma membrane interactions with the cytoskeleton and cell wall. Annual Review of Cell Biology 13:697–743.

Fox, G. E., Stackebrandt, E., Hespell, R. B., Gibson, J., Maniloff, J., Dyer, T. A., Wolfe, R. S., Balch, W. E., Tanner, R. S., Magrum, L. J., Zablen, L. B., Blakemore, R., Gupta, R., Bonen, L., Lewis, B. J., Stahl, D. A., Luehrsen, K. R., Chen, K. N., and Woese, C. R. (1980) The phylogeny of prokaryotes. Science 209:457–463.

Frankel, J. (1989) Pattern Formation: Ciliate Studies and Models. Oxford University Press, New York, Oxford.

Frankel, J. (1990) Positional order and cellular handedness. Journal of Cell Science *97*:205–211.

Frankel, J. (1992) Positional information in cells and organisms. Trends in Cell Biology *1*:256–260.

Frankel, J. (1997) Is spatial pattern formation homologous in unicellular and multicellular organisms? In: Physical Theory in Biology: Foundations and Explorations (C. J. Lumsden, W. A. Brandts, and L. E. H. Trainor, eds.) 245–262. World Scientific, Singapore, New Jersey, London and Hong Kong.

Frankel, J. and Whiteley, A. H. (1993) Vance Tartar: a unique biologist. Journal of Eukaryotic Microbiology *40*:1–9.

Fulton, A. B. (1981) How do eukaryotic cells construct their cytoarchitecture? Cell *24*:4–5.

Gesteland, R. F., Cech, T. R. and Atkins, J. F. (eds.) (1999) The RNA World, 2nd ed. Cold Spring Harbor Laboratory Press, Cold Spring Harbor, New York.

Gilbert, S. F. (1991) Cytoplasmic action in development. Quarterly Review of Biology *66*:309–316.

Gilbert, S. F., Opitz, J. M., and Raff, R. A. (1996) Resynthesizing evolutionary and developmental biology. Developmental Biology *173*:357–372.

Gilbert, W. (1986) Origin of life: the RNA world. Nature *319*:618.

Gilbert, W. and de Souza, S. J. (1999) Introns and the RNA World. In: The RNA World, 2nd ed. (R. F. Gesteland, T. R. Cech, and J. F. Atkins, eds.). Cold Spring Harbor Laboratory Press, Cold Spring Harbor, New York.

Gillin, F. D., Reiner, D. S., and McCaffery, J. M. (1996) Cell biology of the primitive eukaryote *Giardia lamblia*. Annual Reviews of Microbiology *50*: 679–705.

Gogarten, J. P., Hilario, E. and Olendzenski, L. (1996) Gene duplications and horizontal gene transfer during early evolution. In: Evolution of Microbial Life (D. Mc L. Roberts, P. Sharp, G. Alderson and M. Collins, eds.). Society for General Microbiology Symposium *54*:266–292.

Gogarten, J. P., Kiback, H., Dittrich, P., Taiz, L., Bowman, E. J., Bowman, B. J., Manolson, M. F., Poole, R. J., Date, T., Oshima, T., Konishi, J., Denda, K., and Yoshida, M. (1989) Evolution of the vacuolar H^+-ATPase: implications for the origin of eukaryotes. Proceedings of the National Academy of Sciences USA *86*:6661–6665.

Gold, T. (1992) The deep, hot biosphere. Proceedings of the National Academy of Sciences USA *89*:6045-6049.

Goodsell, D. S. (1992) A look inside the living cell. American Scientist Sept.-Oct., 457–465.

Goodwin, B. (1993) Development as a robust natural process. In: Thinking about Biology (W. Stein and F. J. Varela, eds.) 123–148. SFI Studies in the Sciences of Complexity, Lecture Notes Vol. III, Addison-Wesley.

Goodwin, B. (1986) What are the causes of morphogenesis? BioEssays *3*:32–36.

Goodwin, B. (1994) How the Leopard Changed its Spots: The Evolution of Complexity. Simon and Schuster, New York.

Goodwin, B. (1997) General dynamics of morphogenesis. In: Physical Theory in Biology: Foundations and Explorations (C. J. Lumsden, W. A. Brandts, and L. E. H. Trainor, eds.) 187–207. World Scientific, Singapore, New Jersey, London and Hong Kong.

Goodwin, B., Sibatani, A., and Webster, G. eds. (1989) Dynamic Structures in Biology. Edinburgh University Press, Edinburgh.

Goodwin, B. C. and Trainor, L. E. H. (1985) Tip and whorl morphogenesis in *Acetabularia* by calcium-regulated strain fields. Journal of Theoretical Biology *117*:79-106.

Gould, S. J. (1982) Darwinism and the expansion of evolutionary theory. Science *216*: 380–387.

Gould, S. J. (1989) Through a lens, darkly. Natural History *April*:16-24.

Gould, S. J. (1993) Evolution of organisms. In: The Logic of Life—The Challenge of Integrative Physiology (C. A. R. Boyd and D. Noble, eds.). 15-42. Oxford University Press, New York and Oxford.

Gould, S. J. (1994) The evolution of life on the earth. Scientific American, October, pp 85–91.

Gould, S. J. (1996a) Microcosmos. Natural History, March, 22–68 [*sic*].

Gould, S. J. (1996b) Full House: The Spread of Excellence from Plato to Darwin. Crown Publishers, New York.

Gould, S. J. (1997) The exaptive excellence of spandrels as a term and prototype. Proceedings of the National Academy of Sciences U.S.A. *94*;10750-10755.

Gould, S. J. and Eldredge, N. (1993) Punctuated equilibrium comes of age. Nature *366*:223–227.

Gow, N. A. R. and Gadd, G. M. (eds., 1995) The Growing Fungus. Chapman and Hall, London, Glasgow, etc.

Gray, M. W. and Doolittle, W. F. (1982) Has the endosymbiont hypothesis been proven? Microbiological Reviews *46*:1–43.

Gray, M. W. and Spencer, D. F. (1996) Organellar evolution. In Evolution of Microbial Life (D. McL. Roberts, P. Sharp, G. Alderson, and M. Collins, eds.). Society for General Microbiology Symposium *54*:109–126.

Gray, M. W., Burger, G. and Lang, B. F. (1999) Mitochondrial evolution. Science *283*:1476–1481.

Grebecki, A. (1994) Membrane and cytoskeleton flow in motile cells, with emphasis on the contribution of free-living amoebas. International Review of Cytology *148*:37–79.

Grimes, G. W. (1990) Inheritance patterns in ciliated protozoa. In: Cytoplasmic Organization Systems—A Primer in Developmental Biology (G. M. Malacinski, ed.). Vol IV, 23–43. McGraw-Hill, New York.

Grimes, G. W. and Aufderheide, K. J. (1991) Cellular Aspects of Pattern Formation: The Problem of Assembly. Karger, Basel, München, Paris etc.

Gupta, R. S. (1998) Protein phylogenies and signature sequences: A reappraisal of evolutionary relationships among archaebacteria, eubacteria and eukaryotes. Microbiology and Molecular Biology Reviews *62*:1435–1491.

Han, T. M. and Runnegar, B. (1992) Megascopic eukaryotic algae from the 2.1 billion-year-old Negaunee iron-formation, Michigan. Science *257*:232–235.

Harold, F. M. (1977) Ion currents and physiological functions in microorganisms. Annual Review of Microbiology *31*:181–203.

Harold, F. M. (1978) The 1978 Nobel prize in chemistry. Science *202*:1174–1176.

Harold, F. M. (1986) The Vital Force: A Study of Bioenergetics. W. H. Freeman, New York.

Harold, F. M. (1990) To shape a cell: an inquiry into the causes of morphogenesis of microorganisms. Microbiological Reviews *54*:381–431.

Harold, F. M. (1994) Ionic and electrical dimensions of hyphal growth. In: The Mycota, Vol. I (J. G. H. Wessels and F. Meinhardt, eds.) 89–109. Springer Verlag, Berlin, Heidelberg, New York and Tokyo.

Harold, F. M. (1995) From morphogenes to morphogenesis. Microbiology *141*:2765–2778.

Harold, F. M. (1997) How hyphae grow: morphogenesis explained? Protoplasma *197*:137–147.

Harold, F. M. (1999) In pursuit of the whole hypa. Fungal Genetics and Biology *27*:128–133.

Harold, F. M. and Maloney, P. C. (1996) Energy transduction by ion currents. In: Neidhardt et al., op. cit., 283–306.

Harold, R. L., Money, N. P. and Harold, F. M. (1996) Growth and morphogenesis in *Saprolegnia ferax*: is turgor required? Protoplasma *191*:105–114.

Harrison, L. G. (1993) Kinetic Theory of Living Pattern. Cambridge University Press.

Hayes, J. M. (1996) The earliest memories of life on earth. Nature *384*:21–22.

Heath, I. B., ed. (1990) Tip Growth in Plant and Fungal Cells. Academic Press, Inc., San Diego, New York, Boston, etc.

Ho, M. W. (1993) The Rainbow and the Worm: The Physics of Organisms. World Scientific, Singapore.

Ho, M. W. (1994) What is (Schrödinger's) negentropy? In: What is Controlling Life? (E. Gnaiger, F. N. Gellerich, and M. Wyss, eds.) Innsbruck University Press, 50–61.

Höltje, J. V. (1996) A hypothetical holoenzyme involved in the replication of the murein sacculus of *Escherichia coli*. Microbiology *142*:1911–1918.

Höltje, J. V. (1998) Growth of the stress-bearing and shape-maintaining murein sacculus of *Escherichia coli*. Microbiology and Molecular Biology Reviews 62:181–203.

Holzman, D. (1999) More wondering about Martian microbes, life's origins. ASM News 65:393–395.

Horgan, J. (1991) In the beginning . . . Scientific American, Feb., pp 17–123.

Hunter, G. K. (1996) Is biology reducible to chemistry? Perspectives in Biology and Medicine 40:130–138.

Hyde, G. J. and Heath, I. B. (1997) Ca^{2+} gradients in hyphae and branches of *Saprolegnia ferax*. Fungal Genetics and Biology 21:238–251.

Hyman, A. A. and Karsenti, E. (1996) Morphogenetic properties of microtubules and mitotic spindle assembly. Cell 84:401–410.

Ingber, D. E. (1993) Cellular tensegrity: defining new rules of biological design that govern the cytoskeleton. Journal of Cell Science 104:613–627.

Ingber, D. E. (1998) The architecture of life. Scientific American, January 1998, 48–57.

Iwabe, N., Kuma, K. I., Hasegawa, M., Osawa, S. and Miyata, T. (1989) Evolutionary relationship of archaebacteria, eubacteria and eukaryotes inferred from phylogenetic trees of duplicated genes. Proceedings of the National Academy of Sciences USA 86:9355–9359.

Jablonka, E. and Lamb, M. J. (1995) Epigenetic Inheritance and Evolution: The Lamarckian Dimension. Oxford University Press, Oxford, New York, Tokyo.

Jacob, F. (1973) The Logic of Life: A History of Heredity. Pantheon Books, New York.

Jacob, F. (1982) The Possible and the Actual. Pantheon Books, New York.

Jacob, F., Brenner, S. and Cuzin, F. (1963) On the regulation of DNA synthesis in bacteria. Cold Spring Harbor Symposia on Quantitative Biology 28:329–347.

Jaffe, L. F. (1981) The role of ionic currents in establishing developmental patterns. Philosophical Transactions of the Royal Society of London, Series B, 295:553–566.

Jarrell, K. F., Bayley, D. P. and Kostyukuva, A. S. (1996) The archaeal flagellum: a unique motility structure. Journal of Bacteriology 178:5057–5064.

Jeffares, D. C., Poole, A. M., and Penny, D. (1998) Relics from the RNA world. Journal of Molecular Evolution 46:18–36.

John, P. and Whatley, F. R. (1975) *Paracoccus denitrificans* and the evolutionary origin of the mitochondrion. Nature 254:495–498.

Joyce, G. F. (1989) RNA evolution and the origin of life. Nature 338:217–224.

Judson, H. F. (1979) The Eighth Day of Creation. Simon and Schuster, New York. (Reprinted and expanded 1996, Cold Spring Harbor Laboratory Press, New York.)

Kakinuma, Y. (1998) Inorganic cation transport and energy transduction in *Enterococcus hirae* and other streptococci. Microbiology and Molecular Biology Reviews *62*:1021–1045.

Kandler, O. and König, H. (1993) Cell envelopes of Archaea: Structure and chemistry. In: The Biochemistry of Archaea (M. Kates, D. J. Kushner and A. T. Matheson, eds.) Elsevier, Amsterdam, etc., 223–259.

Katz, L. A. (1996) Transkingdom transfer of the phosphoglucose isomerase gene. Journal of Molecular Evolution *43*:453–459.

Katz, M. (1986) Templets and the Explanation of Complex Patterns. Cambridge University Press, Cambridge.

Kauffman, S. (1993) The Origins of Order: Self-Organization and Selection in Evolution. Oxford University Press, New York and Oxford.

Kauffman, S. (1995) At Home in the Universe: The Search for Laws of Self-Organization and Complexity. Oxford University Press, New York and Oxford.

Keefe. A. D. and Miller, S. L. (1995) Are polyphosphates or phosphate esters prebiotic reagents? Journal of Molecular Evolution *41*:693–702.

Keeling, P. (1998) A kingdom's progress: Archezoa and the origin of eukaryotes. BioEssays *20*:87–95.

Kellenberger, E. (1990) Form determination of the heads of bacteriophages. European Journal Biochemistry *190*:233–248.

Kidwell, M. G. (1993) Lateral transfer in natural populations of eukaryotes. Annual Review of Genetics *27*:235–256.

Kirschner, M. (1992) Evolution of the cell. In: Molds, Molecules and Metazoa (P. R. Grant and H. S. Horn, eds.). 99–126. Princeton University Press, Princeton.

Kirschner, M. and Gerhart, J. (1998) Evolvability. Proceedings of the National Academy of Sciences U.S.A. *95*:8420–8427.

Knoll, A. H. (1992) The early evolution of eukaryotes: A geological perspective. Science *256*:622–627.

Knoll, A. H. (1999) A new moleuclar window on early life. Science *285*:1025–1026.

Koch, A. L. (1985) How bacteria grow and divide in spite of internal hydrostatic pressure. Canadian Journal of Microbiology *31*:1071–1083.

Koch, A. L. (1988) Biophysics of bacterial walls viewed as a stress-bearing fabric. Microbiological Reviews *52*:337–353.

Koch, A. L. (1992) Differences in the formation of poles of *Enterococcus* and *Bacillus*. Journal of Theoretical Biology *154*:205–217.

Koch, A. L. (1994) Development and diversification of the last universal ancestor. Journal of Theoretical Biology *168*:269–280.

Koch, A. L. (1995) Bacterial Growth and Form. Chapman and Hall, New York.

Koch, A. L. and Höltje, J. V. (1995) A physical basis for the precise location of the division site of rod-shaped bacteria: the Central Stress Model. Microbiology *141*:3171–3180.

Koch, A. L. and Schmidt, T. M. (1991) The first cellular bioenergetic process: primitive generation of a proton motive force. Journal of Molecular Evolution *33*:297–304.

Koch, A. L., Higgins, M. L. and Doyle, R. J. (1981) Surface-tension-like forces determine bacterial shapes. Journal of General Microbiology *123*:151–161.

Koshland, D. E., Jr. (1977) A response regulator model in a simple sensory system. Science *196*:1055–1063.

Kropf, D. L. (1992) Establishment and expression of cell polarity in fucoid zygotes. Microbiological Reviews *56*:316–339.

Kuhn, T. (1970) The Structure of Scientific Revolutions. Second Edition, University of Chicago Press, Chicago, Ill.

Küppers, B. O. (1990) Information and the Origin of Life. M.I.T. Press, Boston.

Lancelle, S. A., Cresti, M. and Hepler, P. K. (1997) Growth inhibition and recovery in freeze substituted *Lilium longiflorum* pollen tubes: structural effects of caffeine. Protoplasma *196*:21–33.

Landman, O. (1991) The inheritance of acquired characteristics. Annual Review of Genetics *25*:1–20.

Langworthy, T. A. and Pond, J. L. (1986) Membranes and lipids of thermophiles. In: Thermophiles: General, Molecular and Applied Microbiology (T. D. Brock, ed.), 107–135, John Wiley and Sons, New York.

Laszlo, E. (1972) Introduction to Systems Philosophy. Gordon and Breach, London.

Lauffenburger, D. A. and Horwitz, A. F. (1996) Cell migration: a physically integrated molecular process. Cell *84*:359–369.

Lawrence, J. G. and Ochman, H. (1998) Molecular archaeology of the *E. coli* genome. Proceedings of the National Academy of Sciences U.S.A. *95*:9413–9417.

Lawrence, P. A. and Struhl, G. (1996) Morphogens, compartments, and patterns: lessons from *Drosophila*? Cell *85*:951–961.

Lazcano, A. and Miller, S. L. (1996) The origin and early evolution of life: prebiotic chemistry, the pre-RNA world, and time. Cell *85*:793–798.

Lederberg, J. (1966) In: Current Topics in Developmental Biology (A. A. Moscona, and A. Monroy, eds.), *1*: 10.

Lee, J., Ishihara, A. and Jacobson, K. (1993) How do cells move along surfaces? Trends in Cell Biology *2*:366–370.

Lehninger, A. L., Nelson, D. L. and Cox, M. M. (1993) Principles of Biochemistry, 2nd ed. Worth Publishers, New York, U.S.A.

Lenton, T. M. (1998) Gaia and natural selection. Nature *394*:439–447.

Leopold, A. (1949) A Sand County Almanac, and Sketches Here and There. Oxford University Press, New York.

Lewin, R. (1992) Complexity: Life at the Edge of Chaos. Macmillan Publishing Co., New York.

Lifson, F. (1997) On the crucial stages in the origin of animate matter. Journal of Molecular Evolution *44*:1–8.

Lutkenhaus, J. and Addinall, S. G. (1997) Bacterial cell division and the Z ring. Annual Review of Biochemistry *66*:93–116.

Lutkenhaus, J. and Mukherjee, A. (1996) Cell division. In: Neidhardt et al., op. cit, 1615–1626.

Macnab, R. M. (1996) Flagella and motility. In: Neidhardt et al. (op. cit.), pp 123–145.

Maddock, J. R., Alley M. R. K., and Shapiro, L. (1993) Polarized cells, polarized actions. Journal of Bacteriology *175*:7125–7129.

Maden, B. E. H. (1995) No soup for starters? autotrophy and the origins of metabolism. Trends in Biochemical Sciences *20*:337–341.

Maniloff, J. (1996) The minimal cell genome: "On being the right size." Proc. Natl. Acad. Sci. *93*:1004–1006.

Margulis, L. (1970) Origin of Eukaryotic Cells. Yale University Press, New Haven.

Margulis, L. (1992) Biodiversity: molecular biological domains, symbiosis and kingdom origins. BioSystems *27*:39–51.

Margulis, L. (1993) Symbiosis in Cell Evolution, 2nd ed., W. H. Freeman, New York.

Margulis, L. (1996) Archaeal-eubacterial mergers in the origin of Eukarya: phylogenetic classification of life. Proceedings of the National Academy of Sciences USA *93*:1071–1076.

Margulis, L. and Fester, R. (eds.) (1991) Symbiosis as a Source of Evolutionary Innovation: Speciation and Morphogenesis. The MIT Press, Cambridge (Mass), London.

Margulis, L. and Schwartz, K. V. (1998) Five Kingdoms. An Illustrated Guide to the Phyla of Life on Earth, 3rd ed. W. H. Freeman and Co., San Francisco.

Margulis, L., Corliss, J. O., Melkonian, M., and Chapman, D. J. (eds.) (1990) Handbook of Protoctista. Jones and Bartlett Publishers, Boston.

Martin, W. and Müller, M. (1998) The hydrogen hypothesis for the first eukaryote. Nature *342*:37–41.

Mata, J. and Nurse, P. (1998) Discovering the poles in yeast. Trends in Cell Biology *8*:163–167.

Mathews, C. K. (1993) The cell-bag of enzymes or network of channels? Journal of Bacteriology *175*:6377–6381.

Maynard Smith, J. (1986) The Problems of Life. Oxford University Press, Oxford and New York.

Maynard Smith, J. and Szathmàry, E. (1995) The Major Transitions in Evolution. W. H. Freeman, Oxford, New York.

Mayr, E. (1982) The Growth of Biological Thought: Diversity, Evolution and Inheritance. Harvard University Press, Cambridge and London.

Mayr, E. (1988) Is biology an autonomous science? In: Towards a New Philosophy of Biology, Harvard University Press, Cambridge and London.

Mayr, E. (1990) A natural system of organisms. Nature *348*:491.

Mayr, E. (1997) The objects of selection. Proceedings of the National Academy of Sciences U.S.A. *94*:2091–2094.

Mayr, E. (1998) Two empires or three? Proceedings of the National Academy of Sciences USA *95*:9720–9723.

Medawar, P. (1982) Pluto's Republic. Oxford University Press, New York.

Meinhardt, H. (1982) Models of Biological Pattern Formation. Academic Press, Inc. (London) Ltd., London.

Midgley, M. (1992) Science as Salvation. Routledge, London.

Miller, S. L. and Lazcano, A. (1995) The origin of life: did it occur at high temperatures? Journal of Molecular Evolution *41*:689–692.

Mitchell, P. (1962) Metabolism, transport and morphogenesis: which drives which? Journal of General Microbiology *29*:25–37.

Mitchell, P. (1966) Chemiosmotic coupling in oxidative and photosynthetic phosphorylation. Biological Reviews of the Cambridge Philosophical Society *41*:445–502.

Mitchell, P. (1979) David Keilin's respiratory chain and its chemiosmotic consequences. Science *206*:1148–1159.

Mitchison, T. J. (1995) Evolution of a dynamic cytoskeleton. Philosophical Transactions of the Royal Society of London, Series B, *349*:299–304.

Mittenthal, J. E., Clarke, B., and Levinthal, M. (1993) Designing bacteria. In: Thinking about Biology (W. Stein and F. J. Varela, eds.) Addison Wesley Publishing Co., Reading, MA.

Money, N. P. (1997) Wishful thinking about turgor revisited: the mechanics of fungal growth. Fungal Genetics and Biology *21*:173–187.

Monod, J. (1971) Chance and Necessity. Random House, New York.

Moreira, D. and Lopez Garcia, P. (1998) Symbiosis between methanogenic Archaea and δ-Proteobacteria as the origin of eukaryotes: the syntrophic hypothesis. Journal of Molecular Evolution *47*:513–530.

Morowitz, H. J. (1968) Energy Flow in Biology: Biological Organization as a Problem in Thermal Physics. (Reprinted 1979 by OxBow Press, Woodbridge, CT.)

Morowitz, H. J. (1987) Cosmic Joy and Local Pain: Musings of a Mystic Scientist. Charles Scribner's Sons, New York.

Morowitz, H. J. (1992) Beginnings of Cellular Life. Yale University press, New Haven and London.

Muller, H. J. (1958) The Loom of History. Harper and Brothers, New York.

Mushegian, A. R. and Koonin, E. V. (1996) A minimal gene set for cellular life derived by comparison of completed bacterial genomes. Proc. Natl. Acad. Sci. *93*:10268–10273.

Nanney, D. L. (1977) Molecules and morphologies: the perpetuation of pattern in ciliated protozoa. J. Protozool. *24*:27–35.

Nanninga, N. (1998) Morphogenesis of *Escherichia coli*. Microbiology and Molecular Biology Reviews *62*:110–129.

Nasmyth, K. (1995) Evolution of the cell cycle. Philosophical Transactions of the Royal Society of London, Series B, *349*:271–281.

Neidhardt, F. C. and Savageau, M. A. (1996) Regulation beyond the operon. In: Neidhardt et al., op. cit., 1310–1324.

Neidhardt, F. C. and Umbarger, W. E. (1966) In: Neidhardt et al., op. cit., pp 13–16.

Neidhardt, F. C., Ingraham, J. L. and Schaechter, M. (1990) Physiology of the Bacterial Cell: A Molecular Approach. Sinauer Associates, Sunderland, MA.

Neidhardt, F. C., Curtiss, R. III, Ingraham, J. L., Lin, E. C. C., Low, K. B., Magasanik, B., Reznikoff, W. L., Riley, M., Schaechter, M., and Umbarger, H. E. (eds.: 1996). *Escherichia coli* and *Salmonella*: Cellular and Molecular Biology, Second Edition, ASM Press, Washington.

Neumann, C. and Cohen, S. (1997) Morphogens and pattern formation. BioEssays *19*:721–729.

Newport, J. (1987) Nuclear reconstitution *in vitro*: stages of assembly around protein-free DNA. Cell *48*:205–217.

Nicholls, D. G. and Ferguson, S. J. (1992) Bioenergetics 2. Academic Press Ltd., London.

Nicolis, G. and Prigogine, I. (1989) Exploring Complexity: An Introduction. W. H. Freeman and Co., New York.

Nijhout, N. F. (1990) Metaphors and the role of genes in development. BioEssays *12*:441–446.

Norris, V., Turnock, G. and Sigbee, D. (1996) The *Escherichia coli* enzoskeleton. Molecular Microbiology *19*:197–204.

Nüsslein-Volhardt, C. (1996) Gradients that organize embryo development. Scientific American, August, 54–61.

Ochman, H. and Lawrence, J. G. (1996) Phylogenetics and the amelioration of bacterial genomes. In: Neidhardt et al., op. cit., pp 2627–2637.

Ohta, T. and Kreitman, M. The neutralist-selectionist debate. BioEssays *18*: 673–683.

Olsen, G. J. and Woese, C. R. (1997) Archaeal genomics: an overview. Cell *89*:991–994.

Orgel, L. E. (1994) The origin of life on the earth. Scientific American, October *271*:76–83.

Orgel, L. E. (1998) The origin of life: a review of facts and speculations. Trends in Biochemical Sciences *23*:491–495.

Pace, N. R. (1991) Origin of life—facing up to the physical setting. Cell *65*: 531–533.

Pace, N. R. (1997) A molecular view of microbial diversity and the biosphere. Science *276*:734–740.

Park, J. T. (1996) The murein sacculus. In: Neidhardt et al., op. cit, pp 48–57.

Parkinson, J. S. (1993) Signal transduction schemes of bacteria. Cell *73*:857–871.

Perutz, M. (1986) A new view of Darwinism. New Scientist *112*, October, pp 36–38.

Perutz, M. (1987) Physics and the riddle of life. Nature. *326*:555–558.

Pettijohn, D. E. (1996) The nucleoid. In Neidhardt et al., op. cit., pp 158–166.

Pirie, N. W. (1938) The meaninglessness of the terms "life" and "living." In: Perspectives in Biochemistry (J. Needham and D. Green, eds.), pp 11–22, Cambridge, London.

Pittendrigh, C. S. (1993) Temporal organization: reflections of a Darwinian clock watcher. Annual Review of Physiology *55*:17–54.

Polanyi, M. (1968) Life's irreducible structure. Science *160*:1308–1312.

Poundstone, W. (1985) The Recursive Universe. William Morrow and Company, New York.

Preer, J. R. (1993) Unconventional genetic systems. Perspectives in Biology and Medicine *36*:395–419.

Prigogine, I. and Stengers, I. (1984) Order out of Chaos: Man's New Dialogue with Nature. Bantam Books, Toronto, New York, London and Sydney.

Raff, R. A. (1996) The Shape of Life: Genes, Development and the Evolution of Animal Form. University of Chicago Press, Chicago and London.

Raskin, D. M. and De Boer, P. A. J. (1999) Rapid pole to pole oscillation of a protein required for directing division to the middle of *E. coli*. Proceedings of the National Academy of Sciences USA *96*:4971–4976.

Rensberger, B. (1997) Life Itself: Exploring the Realm of the Living Cell. Oxford University Press, New York, etc.

Riedl, R. (1978) Order in Living Organisms. John Wiley and Sons, Chichester, New York, Brisbane, Toronto.

Riley, M. (1993) Functions of the gene products of *Escherichia coli*. Microbiological Reviews *57*:862–952.

Riley, M. and Labedan, B. (1996) *Escherichia coli* gene products: physiological functions and common ancestries. In: Neidhardt et al., op. cit., pp 2118–2202.

Riquelme, M., Reynaga-Peña, C. G., Gierz, G., and Bartnicki-Garcia, S.(1998) What determines growth direction in fungal hyphae? Fungal Genetics and Biology *24*:101–109.

Ritter, W. E. (1919) Cited in Becker, M. O., Organismic biology. In: Encyclopedia of Philosophy (P. Edwards, ed.), p 549. MacMillan and Co., New York, 1967.

Roberson, R. W. and Fuller, M. S. (1988) Ultrastructural aspects of the hyphal tip of *Sclerotium rolfsii* preserved by freeze substitution. Protoplasm *146*: 143–149.

Roemer, T., Vallier, L. G. and Snyder, M.(1996) Selection of polarized growth sites in yeast. Trends in Cell Biology *6*:434–441.

Rosen, R. (1985) Organisms as causal systems which are not mechanisms: an essay into the nature of complexity. In: Theoretical Biology and Complexity, 165–204. Academic Press, Orlando etc.

Rosen, R. (1991) Life Itself: A Comprehensive Inquiry into the Nature, Origin and Fabrication of Life. Columbia University Press, New York.

Rosenberg, A. (1985) The Structure of Biological Science. Cambridge University Press, Cambridge.

Rothfield, L. R. (1994) Bacterial chromosome segregation. Cell *77*:963–966.

Rothfield, L. R. and Zhao, C. R. (1996) How do bacteria decide where to divide? Cell *84*:183–186.

Runnegar, B. (1994) Proterozoic eukaryotes: evidence from biology and geology. In: Early Life on Earth (S. Bengtson, ed.), pp 287–297. Columbia University Press, New York.

Russell, E. S. (1916) Form and Function: A Contribution to the History of Animal Morphology. Murray, London.

Sagan, C., Thompson, W. R., Carlson, R., Gurnett, D. and Hord, C. (1993) A search for life on earth from the Galileo spacecraft. Nature *365*:715–721.

Sapp, J. (1994) Evolution by Association: A History of Symbiosis. Oxford University Press, New York and Oxford.

Sapp, J. (1998) Cytoplasmic heretics. Perspectives in Biology and Medicine *41*:224–240.

Saunders, P. R. (1993) In: Thinking about Biology. (W. Stein and F. J. Varela, eds.) Addison Wesley.

Schoonen, M. A. A., Xu, Y., and Bebie, J. (1999) Energetics and kinetics of the prebiotic synthesis of simple organic acids and amino acids with the FeS-H_2S/FeS$_2$ redox couple as reductant. Origins of Life and Evolution of the Biosphere *29*:5–32.

Schopf, J. W. (1993) Microfossils of the early archaean Apex chert: new evidence of the antiquity of life. Science *260*:640–646.

Schopf, J. W. (1996) Are the oldest fossils cyanobacteria? In: Evolution of Microbial Life (D. McL. Roberts, P. Sharp, G. Alderson, and M. Collins, eds.). Society for General Microbiology Symposium *54*:22–61.

Schrödinger, E. (1944) What is Life? Cambridge University Press. Cambridge.

Schwann, T. (1847) Microscopical Researches into the Accordance in the Structure and Growth of Animals and Plants. English translation by H. Smith. The Sydenham Society, London.

Shapiro, J. A. (1997) Genome organization, natural genetic engineering and adaptive mutation. Trends in Genetics *13*:98–104.

Shapiro, L. and Losick, R. (1997) Protein localization and cell fate in bacteria. Science *276*:712–717.

Shapiro, R. (1986) Origins: A Skeptic's Guide to the Creation of Life on Earth. Summit Books, New York.

Shapiro, R. (1999) Prebiotic cytosine synthesis: A critical analysis and implications for the origin of life. Proceedings of the National Academy of Sciences USA *96*:4396–4401.

Simon, H. (1962) The architecture of complexity. Proceedings of the American Philosophical Society *106*:467–482.

Simpson, G. G. (1963) Biology and the nature of science. Science *134*:81–88.

Smith, C. U. M. (1976) The Problem of Life. Wiley and Sons, New York.

Sogin, M. L. (1994) The origin of eukaryotes and evolution into major kingdoms. In: Early Life on Earth (S. Bengtson, ed.), pp 181–192. Columbia University Press, New York.

Sogin, M. L., Silberman, J. D., Hinkle, G., and Morrison, H. A. (1996) Problems with molecular diversity in the eukarya. In: Evolution of Microbial Life. (D. McL. Roberts, P. Sharp, G. Alderson, and M. Collins, eds.). Society for General Microbiology Symposium *54*:167–184.

Sonneborn, T. M. (1970) Gene action in development. Proceedings of the Royal Society London, Series B, *176*:347–366.

Spiro, P. A., Parkinson, J. S. and Othmer, H. G. (1997) A model of excitation and adaptation in bacterial chemotaxis. Proceedings of the National Academy of Sciences U.S.A. *94*:7263–7268.

Stanier, R. Y. (1970) Some aspects of the biology of cells and their possible evolutionary significance. Symposia of the Society for General Microbiology *20*:1–38.

Stanier, R. Y. and van Niel, C. B. (1962) The concept of a bacterium. Archiv der Mikrobiologie *42*:17–35.

Stanier, R. Y., Ingraham, J. L., Wheelis, M. L., and Painter, P. R. (1986) The Microbial World. Fifth Edition Prentice Hall, Englewood Cliffs, New Jersey.

Stent, G. S. (1968) That was the molecular biology that was. Science *160*:390–395.

Stent, G. S. (1978) Paradoxes of Progress, W. H. Freeman, New York.

Stetter, K. O. (1994) The lesson of archaebacteria. In: Early Life on Earth (S. Bengtson, ed.). Columbia University Press, New York. pp 143–152.

Stewart, I. and Golubitski, M. (1992) Fearful Symmetry. Blackwell, New York.

Stock, J. B. and Surette, M. G. (1996) Chemotaxis. In: Neidhardt et al., op. cit., pp 1103–1129.

Syvanen, M. (1994) Horizontal gene transfer: Evidence and possible consequences. Annual Review of Genetics *28*:237–261.

Szathmàry, E. and Maynard Smith, J. (1995) The major evolutionary transitions. Nature *374*:227–232.

Taylor, L. P. and Hepler, P. K. (1997) Pollen germination and tube growth. Annual Review of Plant Physiology and Plant Molecular Biology *48*:461–491.

Theriot, J. A. (1996) Worm sperm and advances in cell locomotion. Cell *84*: 1–4.

Thomas, L. (1974) The Lives of a Cell. Bantam Books, Toronto, New York, etc.

Thompson, D. W. (1961) On Growth and Form. Abridged edition (J. T. Bonner, ed.). Cambridge University Press, London.

Tuchman, B. (1981) Practicing History. Alfred A. Knopf, New York.

Turing, A. M. (1952) The chemical basis of morphogenesis. Philosophical Transactions of the Royal Society of London, Series B, *237*:37–72.

Varela, F. G., Maturana, H. R. and Uribe, R. (1974) Autopoiesis: The organization of living systems, its characterization and a model. BioSystems *5*: 187–196.

Wächtershäuser, G. (1988) Before enzymes and templates: theory of surface metabolism. Microbiological Reviews *52*:452–484.

Wächtershäuser, G. (1992) Groundworks for an evolutionary biochemistry: the iron-sulfur world. Progress in Biophysics and Molecular Biology *58*:85–201.

Wächtershäuser, G. (1994) Life in a ligand sphere. Proceedings of the National Academy of Sciences USA *91*:4283–4287.

Wächtershäuser, G. (1997) The origin of life and its methodological challenge. Journal of Theoretical Biology *187*:483–494.

Waddington, C. H. (1957) The Strategy of the Genes. George, Allen & Unwin, London.

Weaver, W. (1948) Science and complexity. American Scientist *36*:536–544.

Weber, B. H. and Depew, D. J. (1996) Natural selection and self-organization: Dynamical models as clues to a new evolutionary synthesis. Biology and Philosophy *11*:33–65.

Weber, B. H., Depew, D. J. and Smith, J. D. (eds.) (1988) Entropy, Information and Evolution. MIT Press, Cambridge, MA.

Webster, G. and Goodwin, B. (1981) History and structure in biology. Perspectives in Biology and Medicine *25*:26–39.

Webster, G. and Goodwin, B. (1996) Form and Transformation: Generative and Relational Principles in Biology. Cambridge University Press, Cambridge.

Wessels, J. G. H. and Meinhardt, F., eds. (1994) The Mycota. Vol I; Growth, Differentiation and Sexuality. Springer Verlag, Berlin, Heidelberg, etc.

Whatley, J. (1993) Endosymbiotic origin of chloroplasts. International Review of Cytology *144*:259–299.

Wheeler, R. T. and Shapiro, L. (1997) Bacterial chromosome segregation: Is there a mitotic apparatus? Cell *88*:577–579.

Wheelis, M. L., Kandler, O., and Woese, C. R. (1992) On the nature of global classification. Proceedings of the National Academy of Sciences USA *89*: 2930–2934.

Whitaker, R. H. (1959) On the broad classification of organisms. Quarterly Review of Biology *34*:210–226.

White, D. (1995) The Physiology and Biochemistry of Prokaryotes. Oxford University Press, New York.

Whitman, W. B., Coleman, D. C., and Wiebe, W. J. (1998) Prokaryotes—the unseen majority. Proceedings of the National Academy of Sciences USA *95*:6578–6583.

Wicken, J. S. (1987) Evolution, Thermodynamics and Information: Extending the Darwinian Program. Oxford University Press, New York.

Wilkins, A. (1996) Are there Kuhnian revolutions in biology? BioEssays *18*: 695–696.

Woese, C. R. (1987) Bacterial evolution. Microbiological Reviews *51*:221–271.

Woese, C. R. (1998a) The universal ancestor. Proceedings of the National Academy of Sciences USA *95*:6854–6859.

Woese, C. R. (1998b) Default taxonomy: Ernst Mayr's view of the microbial world. Proceedings of the National Academy of Sciences USA *95*:11043–11046.

Woese, C. R. and Fox, G. E. (1977) Phylogenetic structure of the prokaryotic domain: the primary kingdoms. Proceedings of the National Academy of Sciences USA *74*:5088–5900.

Woese, C. R., Kandler, O., and Wheelis, M. L. (1990) Towards a natural system of organisms: proposal for the domains Archaea, Bacteria and Eucarya. Proceedings of the National Academy of Sciences USA *87*:4576–4579.

Woodger, J. H. (1929, reissued 1967) Biological Principles: A Critical Study. Routledge & Kegan Paul Ltd., London.

Zubay, G. (1996) Origins of Life on the Earth and in the Cosmos. Wm. C. Brown Publishers, Dubuque, Iowa.

GLOSSARY

Actin The protein that makes up microfilaments.

Adaptation In evolution, any change in the structure or functioning of an organism that makes it better suited to its environment.

Adenosine diphosphate (ADP) See adenosine triphosphate.

Adenosine triphosphate (ATP) A nucleotide that is of fundamental importance as a carrier of energy in all organisms. It is composed of adenine, ribose and three phosphoryl groups; successive release of phosphate gives rise to *adenosine diphosphate* (ADP) and *adenosine monophosphate* (AMP).

ADP See adenosine triphosphate.

Allele One of two or more alternate states of a gene that typically arise by mutation.

Allostery Shape-change. In biochemistry, the change in shape of a protein resulting from the binding of a small molecule.

Amino acids Small, water-soluble organic compounds that possess both a carboxyl group ($-COOH$) and an amino group ($-NH_2$). Proteins are polymers made up of a characteristic set of twenty amino acids.

Apical growth A common mode of biological growth in which all new material is deposited at a tip or apex. Examples include fungal hyphae, neurons and plant roots.

Archaeon An organism classified in the prokaryotic domain *Archaea*, which contains many bacteria that live in extreme environments.

Archean eon The period from the formation of the planet to 2.5 billion years ago.

Archezoa A kingdom-level division of the protists whose members lack mitochondria and chloroplasts, and which may have diverged from the eukaryotic stem prior to the acquisition of endosymbionts.

ATP See adenosine triphosphate.

ATP synthase A complex enzyme associated with the membranes of organelles and bacteria, that mediates ATP synthesis during oxidative phosphorylation and photosynthetic phosphorylation.

Autopoiesis The capacity of living organisms to make and reproduce themselves autonomously.

Bacteria An informal, traditional term that embraces all prokaryotes. See also Archaea, Eubacteria.

Bacteriophage ('phage) A virus that is parasitic on bacteria.

Biopoiesis The development of life from non-living matter; synonymous with the origin of life.

Brownian motion The continuous random movements of microscopic particles when suspended in a fluid medium, produced by bombardment with molecules of the medium.

Catalyst A substance that increases the rate of a chemical reaction without itself undergoing permanent chemical change. Enzymes are the chief catalysts in biochemical reactions.

Chemolithotroph A bacterium that lives by extracting energy from inorganic chemical reactions.

Chemoreceptor A receptor that detects the presence of a particular set of molecules.

Chemotaxis Movement of a cell in response to a chemical stimulus, towards attractants and away from repellants.

Chitin A tough, resistant polysaccharide that forms the cell walls of fungi and the exoskeleton of insects. It is a polymer of N-acetylglucosamine.

Chloroplast An organelle, found in the cytoplasm of many eukaryotic cells, that contains chlorophyll and carries out photosynthesis.

Chromosome A thread-like structure found in cell nuclei, which carries genes in linear array.

Ciliate A protozoan whose surface is studded with cilia, which serve as the organs of motility.

Cell The structural and functional unit of living organisms. Cell size varies, but most are microscopic (0.001 to 0.1 mm). Many organisms consist of but a single cell (bacteria and most protists), others are multicellular (fungi, plants, and animals).

Cell wall The strong and relatively rigid envelope of many cells, external to the plasma membrane. Cells of plants, fungi, many protists, and most bacteria are walled; animal cells are not.

Cellulose A polymer of glucose characteristically found in eukaryotic cell walls, especially those of plants and certain protists.

Cenancestor See Last common ancestor.

Centrosome A region of the cytoplasm adjacent to the nucleus, which often harbors centrioles and serves as the organizing center for cytoplasmic microtubules.

Cilium See flagellum.

Coding In biology, the procedure by which the sequence of nucleotides in a gene determines the sequence of amino acids in a protein. See Codon.

Codon A triplet of nucleotide bases within a molecule of DNA or mRNA that specifies a particular amino acid.

Complexity The condition of a system whose behavior is not easily deduced from the properties of its parts.

Cortex In cell biology, the cytoplasmic layer that lies beneath the plasma membrane.

Cyanobacteria A division of the Eubacteria, characterized by photosynthesis of the same kind as that of chloroplasts.

Cytokinesis The process by which a cell divides into two (or more) separate daughter cells.

Cytology The study of cell structure, chiefly with the aid of microscopes.

Cytoplasm (Protoplasm) The jelly-like material surrounding the cells nucleus.

Cytoskeleton A network of microscopic filaments and tubules within the cytoplasm of eukaryotic cells that is involved in motility, secretion, and other functions.

Cytosol The fluid phase of cytoplasm, exclusive of organelles.

Deoxyribonucleic acid See DNA.

DNA (Deoxyribonucleic acid) A nucleic acid composed of two intertwined polynucleotide chains; the sugar is deoxyribose. DNA is the genetic material of all cells and of many viruses.

DNA polymerase An enzyme that catalyzes the synthesis of DNA from its nucleotide precursors, typically using an existing strand of DNA as the template. The agent of gene replication.

Emergence The appearance of new properties in a system, that were not present nor easily predictable from the properties of the components.

Endergonic A biochemical reaction that requires energy, and will not proceed unless an external energy source is present. For example, most biosynthetic processes.

Endomembranes A generic term for membranes within the cytoplasm; examples include the endoplasmic reticulum and the Golgi apparatus of eukaryotic cells.

Endosymbiosis A symbiosis in which one of the partners (the endosymbiont) resides within the cytoplasm of the other.

Energy The capacity to do work. In biology, the capacity to drive processes that do not occur spontaneously, such as the synthesis of complex molecules.

Entropy A portion of the energy of a system that is unavailable for work. In a wider sense, a measure of the system's disorder: as disorder increases, so does the entropy.

Enzyme A protein that acts as a catalyst in biochemical reactions.

Epigenesis The approximately stepwise process by which genetic information, modified by environmental factors, is translated into the substance and behavior of an organism.

Epigenetic inheritance Refers to heritable features not encoded in the nucleotide sequence of genes.

Epigenetic landscape The circumstances that govern and constrain the expression of genetic information, so as to channel it into a particular developmental pathway.

Eubacterium An organism classified in the prokaryotic domain Eubacteria, which contains most of the familiar bacteria.

Eukarya (Eucarya) The domain of life that contains all eukaryotes, whether unicellular or multicellular.

Eukaryotes (Eucaryotes) Organisms whose genetic material is enclosed in a true, membrane-bound nucleus. Eukaryotic cells also typically have a cytoskeleton, internal membranes and organelles.

Exergonic A biochemical process that releases energy. For example, respiration.

Exocytosis A mode of secretion in which substances are enclosed within a vesicle that fuses with the plasma membrane and discharges its contents to the outside.

Exon A nucleotide sequence in a gene that codes for part or all of the gene product. Exons are sometimes separated by noncoding sequences called introns.

Fermentation A biochemical pathway that breaks down organic substances in the absence of oxygen, with production of metabolites and usable energy.

Fibroblast An animal cell that secretes fibrous material into the intercellular spaces of connective tissue. Typically flat, elongated or star-shaped.

Field A territory that displays coordinated activity, controlled by the differential distribution of some property or agent. Examples include electrical fields and morphogenetic fields.

Fimbria Hair-like projections commonly seen on the surface of bacterial cells.

Fitness The condition of an organism that is well adapted to its environment, as measured by the ability to reproduce its kind.

Flagellate A motile protist equipped with undulipodia (eukaryotic flagella).

Flagellum/cilium/undulipodium A relatively long, whip-like structure present on the surface of many cells and serving as an organ of motility. In this book, flagellum refers specifically to the bacterial organelles. Eukaryotic flagella (and their shorter cousins, the cilia) have an entirely different structure and are designated *undulipodia*.

Foraminifer A kind of protozoan whose elaborate silica shells are often found fossilized.

Free energy A measure of a system's ability to do work. In biochemistry, the fraction of the energy released by a chemical reaction that is available to do work. The unavailable fraction is the *entropy*.

Gene A unit of heredity, usually consisting of a stretch of DNA that codes for some biological function.

Genome The total gene complement of an organism. More precisely, all the genes carried on a single set of chromosomes.

Genophore General term for structures that carry genes. Examples include the chromosomes of eukaryotes and prokaryotes, also plasmids.

Glycolysis The sequence of reactions by which sugar is converted to ethanol in the absence of oxygen.

Golgi apparatus An assembly of vesicles and membranous tubules found in the cytoplasm of eukaryotic cells, that participates in the secretion of cell products.

Gram-positive, Gram-negative Two classes of bacteria that respond differentially to a certain staining procedure, thanks to differences in envelope structure. Bacteria that take up the stain (Gram-positive) lack the outermost lipopolysaccharide membrane characteristic of the Gram-negative ones.

Growth An increase in the dry weight or volume of an organism through cell enlargement, cell division or both.

Hierarchy An arrangement of component parts into graded ranks, such as the military chain of command or the classification of plants and animals.

Histones A class of proteins that occur in chromosomes in association with DNA.

Holism The doctrine that a system may have properties over and above those of its parts and their organization.

Homeostasis The regulation by an organism of its chemical composition and other aspects of its internal environment.

Hydrogenosome The organelle of energy production in certain protists that inhabit environments devoid of oxygen.

Hydrophoby The property of repelling water. Used here to refer to molecules such as lipids that do not form hydrogen bonds with water.

Hypha A filament of a fungus or oömycete.

Intron (Intervening sequence) A nucleotide sequence within a gene that does not code for the gene product.

Last common ancestor A hypothetical organism ancestral to all three domains of life.

Ligand A molecule that binds to, or is bound by, another molecule.

Lipid Any of a diverse group of organic compounds found in living organisms, that are insoluble in water but soluble in organic solvents such as chloroform or benzene. Examples include fats, oils, steroids, and terpenes.

Lipopolysaccharide A complex molecule consisting of both lipid and carbohydrate moieties, that makes up the outermost envelope of Gram-negative bacteria.

Liposome An artificial vesicle bounded by a lipid bilayer membrane.

Lysosomes Organelles found in the cytoplasm of eukaryotic cells that carry out the breakdown of food particles, and sometimes of cell constituents.

Macromolecule A very large molecule, typically of molecular mass 100,000 daltons and more.

Meiosis (Reductive division) A mode of cell division in eukaryotes that gives rise to four reproductive cells (gametes), each with half the chromosome number of the parent cell.

Membrane A sheet-like tissue that covers, connects or lines cells and organelles. Membranes typically consist of polar lipids and proteins, and may contain additional substances.

Messenger RNA (mRNA) The product of transcription of structural genes. A type of RNA that carries the genetic instructions from DNA into the cytoplasm.

Metabolism The sum of the chemical reactions that take place in a living organism. Compounds that take part in, or are formed by, these reactions are called *metabolites*. A sequence of reactions that generates (or degrades) a metabolite is considered a *metabolic pathway*.

Microfilaments Elements of the cytoskeleton of eukaryotic cells. They are helical threads, 6–7 nanometers in diameter, composed of the protein actin.

Microtubules Elements of the cytoskeleton of eukaryotic cells. Microtubules are cylindrical filaments, about 25 nanometers in diameter, composed of the protein tubulin.

Mitochondrion An organelle of eukaryotic cells that carries out aerobic respiration and serves as the cell's powerhouse.

Mitosis The division of a cell to form two daughter cells, each of which has a nucleus containing the same number and kind of chromosomes as the mother cell.

Modern synthesis The theory of evolution as reformulated in the middle of the twentieth century.

Morphogen A substance whose distribution in a cell or organism supplies a pattern that elicits subsequent developmental differences.

Morphogenesis The development of form and structure in an organism.

Morphogenetic field A territory in which developmental events are subject to a common set of coordinating influences.

Murein See peptidoglycan.

Mutant An organism, or a gene, that has undergone a mutation.

Mutation A heritable change in the genetic material of a cell that may cause it and its descendants to differ from the normal type in appearance or behavior.

Nuclear membrane The membrane that delimits the nucleus and regulates the flow of materials between nucleus and cytoplasm. By definition, eukaryotic nuclei are bounded by a membrane, prokaryotic ones are not.

Nucleic acid A large and complex biological molecule consisting of a chain of nucleotides. There are two types, deoxyribonucleic acid (DNA) and ribonucleic acid (RNA).

Nucleotides The basic building blocks of nucleic acids. Each nucleotide consists of a nitrogenous base, a sugar and one or more phosphate groups.

Nucleus/Nucleoid The large, membrane-bound body embedded in the cytoplasm of eukaryotic cells, that contains the genetic material. A nucleoid is the corresponding structure in prokaryotic cells, which is not enclosed in a membrane.

Operon A functionally integrated genetic unit consisting of adjoining genes that encode proteins, together with loci that control their expression.

Order A state in which the components are arranged in a regular, comprehensible, or predictable manner. Various degrees of order are represented by the letters of the alphabet, the arrangement of a keyboard, and the sequence of amino acids in a protein.

Organelle A minute structure within a cell that has a particular function. Examples include the nucleus, mitochondria, and flagella.

Organism An individual living creature, either unicellular or multicellular.

Organization Purposeful or functional order, as in the arrangement of the parts of a bicycle or a skeleton.

Oxidative phosphorylation The chief mechanism of ATP production in aerobic organisms. The enzymatic generation of ATP coupled to the transfer of electrons from a substrate to oxygen.

Peptide One of a large class of organic compounds consisting of two or more amino acids linked by peptide bonds.

Peptidoglycan The characteristic wall polymer of Eubacteria. It consists of repeating units of muramic acid and N-acetyl glucosamine, crosslinked by short chains of amino acids.

Periplasm The narrow space between the plasma membrane and the cell wall.

Plasma membrane (cell membrane) The membrane that forms the outer limit of a cell and regulates the flow of material into the cell and out.

Phagocytosis The process by which cells engulf and digest minute food particles.

Phagotroph An organism that lives by engulfing food particles.

Phospholipid One of a group of lipids that contain both a phosphate group and one or more fatty acids.

Photosynthesis The chemical process by which green plants (and many other organisms) synthesize organic compounds from CO_2 and water by use of the energy of light.

Phototroph An organism that lives by photosynthesis.

Phylogeny The evolutionary descent of an organism, or a group of related organisms.

Plasmid A unit of DNA within the cytoplasm that can exist and replicate independently of the chromosomes.

Polarity The condition of having parts or areas with contrasting properties, such as front and rear.

Polymer A molecule or complex composed of repeating elements (monomers). For example, proteins are polymers of amino acids.

Polynucleotide Any polymer of nucleotides, including DNA and RNA.

Polypeptide A peptide containing more than ten amino acids, commonly a hundred or more. All proteins are polypeptides.

Prokaryotes (Procaryotes) Organisms whose genetic material is not separated from the cytoplasm by a nuclear membrane. More generally, a grade of cellular organization lacking a true nucleus, cytoskeleton and most organelles. Both Eubacteria and Archaea belong to this grade.

Protein A large molecule consisting of one or more polypeptide chains; the molecular mass ranges from 6,000 daltons to the millions.

Proteinoid Polypeptides and related structures formed by subjecting a mixture of amino acids to dry heat.

Proterozoic eon The period from the close of the Archaean to the beginning of the Cambrian, approximately 2.5 to 0.5 billion years ago.

Protist A member of the kingdom Protista, (or Protoctista), which includes the unicellular eukaryotic organisms and some multicellular lineages derived from them.

Protoctist See Protist.

Protoplasm See Cytoplasm.

Pseudopod A temporary outgrowth of certain cells (especially amoebas) that serves as an organ of locomotion or feeding.

Punctuated equilibrium A model of the pattern of evolution that proposes that long periods of relative constancy are punctuated by episodes of rapid change.

Receptor A molecule on the surface of a cell whose function is to detect a particular stimulus and initiate a response.

Reductionism The doctrine that the properties of a complex system can be largely (or even wholly) understood in terms of its simpler parts or components.

Replication The production of an exact copy. Usually employed in reference to the replication of DNA, in which one strand provides a template for the formation of a complementary strand, which is then copied once more to reproduce the original.

Respiration The utilization of oxygen. In cell biology, the oxidative degradation of organic substances with the production of metabolites and energy. Oxygen usually serves as the oxidant, but sulfate or nitrate sometimes take its place in "anaerobic respiration."

Respiratory chain The biochemical basis of respiration. A cascade of proteins and other metabolites that carries electrons from substrates to oxygen in aerobic cells.

Ribonucleic acid. See RNA.

Ribosomes Intracellular organelles that carry out protein synthesis, found in all cells. They are composed of several species of RNA and a number of proteins.

Ribozyme A biological catalyst composed of RNA (unlike enzymes, which are proteins).

RNA (Ribonucleic acid) A nucleic acid made up of polynucleotide chains whose sugar is ribose. Examples include ribosomal RNA, transfer RNA and messenger RNA.

RNA polymerase An enzyme that catalyzes the synthesis of RNA from its nucleotide precursors, typically using an existing strand of DNA or RNA as the template. The agent of gene transcription.

Sequence The order of amino acids in a protein or of nucleotides in a nucleic acid. The determination of that order is referred to as *sequencing*.

Septum A wall that divides a cell into two compartments.

Spandrel An architectural term employed by some evolutionists to describe biological features that do not arise as adaptations.

Spitzenkörper An aggregate of secretory vesicles characteristically found at the tip of growing fungal hyphae.

Steroid Any of a group of lipids derived from a certain phenanthrene derivative that has a nucleus of four rings. Examples include cholesterol and steroid hormones.

Stromatolite A rocky, cushion-like mass, produced by the growth of large populations of cyanobacteria and algae. They are commonly found fossilized, though living ones still occur in favorable locales.

Surface tension The property of a liquid that makes it behave as though its surface were covered with an elastic skin (a result of intermolecular forces). Also, a measure of the work that must be done to enlarge that surface.

Symbiosis Living together: an interaction between individuals of different species that is beneficial to both partners.

Symport/Antiport The coupled transport of two substrates by a single carrier. We speak of symport when both substrates move in the same direction, and of antiport when they move in opposite directions.

System An entity composed of elements that interact, or are related to one another, in some definite manner. A bicycle and a cell are systems; a lump of granite is not.

Templet An object whose structure or configuration supplies a pattern for the assembly of another object.

Thermophile An organism that can live at high temperatures.

Transcription Assembly of an RNA molecule complementary to a stretch of DNA. This is the first step in protein synthesis, and represents the transfer of sequence information from DNA to RNA.

Transfer RNA (tRNA) A small RNA molecule that conducts activated amino acids to the ribosome during protein synthesis.

Translation Assembly of a protein by a ribosome, using messenger RNA to specify the order of the amino acids.

Transposon A mobile genetic element that can become integrated at many different sites along a chromosome.

Tubulin A globular protein that is the chief constituent of microtubules.

Turgor The condition of a walled cell when its cytoplasm is distended by water and exerts pressure on the cell wall.

Undulipodium Eukaryotic flagellum. See flagellum.

Vacuole A membrane-bound space within the cytoplasm of a living cell that is filled with fluid or gas. In plants and some protists, vacuoles occupy much of the cells' volume.

Vectorial Having a direction in space.

Vesicle A small, membrane-bound sac (usually filled with fluid) within the cytoplasm of a living cell.

Virus A particle too small to be seen with the light microscope or to be trapped by filters, but capable of reproduction within a living cell. Viruses have very limited metabolic capacities and are obligatory intracellular parasites.

Work A process that runs counter to the spontaneous direction of events. For example, protein synthesis represents work, but their degradation does not.

Zygote A fertilized egg (more generally, the product of fusion of two gametes), whose development is just beginning.

INDEX

DA